Life in the Rocky Mountains

A Diary of Wanderings on the Sources of the Rivers Missouri, Columbia, and Colorado, 1830-1835

By Warren Angus Ferris

Published by Pantianos Classics

ISBN-13: 978-1545377161

First published in 1843

Contents

Chapter I 6
Chapter II 8
Chapter III 10
Chapter IV 13
Chapter V 16
Chapter VI 18
Chapter VII 21
Chapter VIII 23
Chapter IX 27
Chapter X 31
Chapter XI 35
Chapter XII 38
Chapter XIII 40
Chapter XIV 43
Chapter XV 47
Chapter XVI 50
Chapter XVII 53
Chapter XVIII 56
Chapter XIX 59
Chapter XX 62
Chapter XXI 66
Chapter XXII 69
Chapter XXIII 72

Chapter XXIV ... 76
Chapter XXV ... 79
Chapter XXVI ... 82
Chapter XXVII ... 85
Chapter XXVIII ... 88
Chapter XXIX ... 89
Chapter XXX ... 92
Chapter XXXI ... 94
Chapter XXXII ... 97
Chapter XXXIII ... 100
Chapter XXXIV ... 102
Chapter XXXV ... 105
Chapter XXXVI ... 107
Chapter XXXVII ... 109
Chapter XXXVIII ... 111
Chapter XXXIX ... 113
Chapter XL ... 115
Chapter XLI ... 117
Chapter XLII ... 121
Chapter XLIII ... 124
Chapter XLIV ... 128
Chapter XLV ... 130
Chapter XLVI ... 132
Chapter XLVII ... 135
Chapter XLVIII ... 138
Chapter XLIX ... 140
Chapter L ... 142
Chapter LI ... 144

Chapter LII 146
Chapter LIII 149
Chapter LIV 150
Chapter LV 152
Chapter LVI 154
Chapter LVII 155
Chapter LVIII 156
Chapter LIX 158
Chapter LX 160
Chapter LXI 161
Chapter LXII 163
Chapter LXIII 165
Chapter LXIV 167
Chapter LXV 168
Chapter LXVI 171
Chapter LXVII 172
Chapter LXVIII 174

Supplementary Articles 178

Number 1 - Western Literary Messenger, July 20, 1842 178
Number 2 - Western Literary Messenger, August 17, 1842 179
Number 3 - Dallas Herald, January 11, 1873 181
Number 4 - Dallas Herald, January 27, 1873 184
Number 5 - Dallas Herald, December 14, 1872 189
Number 6 - Dallas Herald, December 14, 1872 191
Number 7 - Dallas Herald, November 30, 1872 192
Number 8 - Dallas Herald, January 4, 1873 194

Verse 196

Chapter I

Westward! Ho! It is the sixteenth of the second month A. D. 1830. and I have joined a trapping, trading, hunting expedition to the Rocky Mountains. Why, I scarcely know, for the motives that induced me to this step were of a mixed complexion, - something like the pepper and salt population of this city of St. Louis. Curiosity, a love of wild adventure, and perhaps also a hope of profit, - for times are hard, and my best coat has a sort of sheepish hang-dog hesitation to encounter fashionable folk - combined to make me look upon the project with an eye of favour. The party consists of some thirty men, mostly Canadians; but a few there are, like myself, from various parts of the Union. Each has some plausible excuse for joining, and the aggregate of disinterestedness would delight the most ghostly saint in the Roman calendar. Engage for money! no, not they; health, and the strong desire of seeing strange lands, of beholding nature in the savage grandeur of her primeval state, - these are the only arguments that could have persuaded such independent and high-minded young fellows to adventure with the American Fur Company in a trip to the mountain wilds of the great west. But they are active, vigorous, resolute, daring, and such are the kind of men the service requires. The Company have no reason to be dissatisfied, nor have they. Everything promises well. No doubt there will be two fortunes apiece for us. Westward! Ho!

All was at last ready, we mounted our mules and horses, and filed away from the Company's warehouse, in fine spirits, and under a fine sky. The day was delightful, and all felt its cheerful influence. We were leaving for many months, - even years - if not forever, the lands and life of civilization, refinement, learning order and law, plunging afar into the savageness of a nomadic, yet not pastoral state of being, and doomed to encounter hunger, thirst, fatigue, exposure, peril, and perhaps sickness, torture, and death. But none of these things were thought of. The light jest was uttered, the merry laugh responded. Hope pictured a bright future for every one, and dangers, hardships, accidents, and disappointments, found no harbour in our anticipations.

The first day's march conducted us through a fertile and cultivated tract of country, to the Missouri river, opposite St. Charles. We crossed the stream in a flat boat, and passing through the village, halted for the night at a farmhouse a few miles beyond. Corn and corn-stalks were purchased for our horses, and corn bread and bacon for ourselves. We did not greatly relish a kind of diet so primitive, neither did we the idea that it was furnished us merely because it was the cheapest that could be obtained. Having ascertained, however, that nothing better was to be had, we magnanimously concluded to accept that instead of the alternative - nothing - and I, at least, made out a hearty supper.

It is not necessary to mention every trifling accident that occurred during our journey through the state of Missouri. Our numbers prevented us from enjoying the comforts of a house to lodge in, and when we could not find room in barns or other outbuildings, we slept on the bosom of mother earth, beneath our own good blankets, and the starry coverlet of heaven. No unpleasant effects resulted from this exposure, and though all unused to a mode of life so purely aboriginal, I even enjoyed it. Sleep more refreshing, and dreams more sweet were never vouchsafed to me than those which waited upon my grassy couch beneath the sky canopy of night. In fair weather nothing could be finer, but a cold driving storm made all the difference in the world. In such an event we arose, took up our beds, and walked - to the nearest door, which we ordered instantly to unfold and yield admittance, on pain of our displeasure. The conscious door trembled at the summons, but never hesitated to obey the mandate, and thereupon we entered and spread ourselves and blankets on the floor, if wet - to dry, if dry - to snore.

On the twenty-first we entered the Eighteen-mile prairie, east of Franklin, beneath a bright sky, and a balmy air. A few miles and the weather changed sadly. A terrible storm set in, which we were obliged to face and brave, for shelter was out of the question. The snow and hail melted and froze again on our hair, eye-brows, and neck-cloths, and we suffered much during almost the whole day from its driving violence. At evening we re-entered the woodlands, and the storm ceased to annoy us. Two days after this, we reached and passed through the village of Franklin, which a pitiless monster was in the act of swallowing up. The river is every year encroaching on the bank that forms the site of the town, and several buildings have already made an aquatic excursion. Others seem preparing to follow. Near the village we met with innumerable flocks of paroquets - the first I had seen in a wild state - whose beautiful plumage of green and gold flashed above us like an atmosphere of gems.

We crossed the Missouri at Arrow-rock ferry on the twenty-fifth, and shortly after overtook a party of fifteen Canadians, who had preceded us a few days from St. Louis, and who were henceforth to be our companions to the end of the journey. The country had already begun to assume a more uncultivated and dreary aspect; plantations were much less frequent, - we were approaching the limits of civilization. We now moved from farm-house to farm-house, remaining at each so long as we could obtain sustenance for ourselves and horses, in order that the condition of the latter might be improved, and to give time for the vegetation, to which their diet would soon have to be restricted, to increase sufficient for the purpose. In the meantime our leisure hours were occupied and amused by the surprising relations of a few of the Canadians who had formerly been to the mountains, and who did not scruple to impose on the credulity of the "mangeris de lard," as they term those who are unacquainted with the wild hap-hazard sort of life peculiar to the remote and desolate regions to which our journey tends. Each of these veterans seemed to have had a "most enormous experience" in mountain adventure, and certainly if their own stories could have been taken for it, they were singly more than a match for any given number of bears or Blackfeet. Some of their narrations were romantic enough, with a possibility of their being true, but the greatest number savoured too much of Munchausenism to gain a moment's belief. I soon found that a current of rude but good natured humour ran through their veins, and that, though quite disposed to quiz, they were by no means disposed to quarrel with us. We easily came to a good understanding together. They told as extravagant yarns as they pleased, and we believed as little as we liked. Both had reason to be pleased with this arrangement, and many an hour I sat and listened to extempore adventures, improvised for the occasion, compared to which those of Colter and Glass, (both of which I had read years before,) were dull and spiritless. One told of coursing an antelope a week without intermission or food, over a spur of the Wind Mountains, and another of riding a grizzly bear, full tilt, through a village of Blackfeet Indians! There was no end to their absurdities.

Chapter II

Messrs. Dripps and Robidoux, who were to be our conductors to the Council Bluffs, overtook us on the fifth, bringing with them an addition to our strength of fifty more - mules! As these our new leaders (not the mules) were noted for anything but a want of energy, we were soon again in motion, and recrossing the Missouri near Mount Vernon, continued our course to a plantation not far from Liberty, the last village on our route, where we remained for two weeks, waiting the arrival of wagons from St Louis, with merchandize for the Indian trade, which from this point has to be conveyed to the mountains on pack horses.
The only incident by which the monotony of our stay was at all relieved, was a stab which one of our men received in a drunken frolic, from a stranger whom he had without doubt insulted. This affair produced at first some little excitement, and even threatened serious consequences. It was soon ascertained, however, that the injury was but slight, and, as the individual wounded was known to be a reckless, impudent quarrelsome fellow, who had beyond question provoked the broil in which he got his hurt, he found but little sympathy, and was forced to put up with the loss of blood and temper his insolence and ill-conduct had brought upon him. This lesson was not entirely lost to him, for it had the effect of amending his manners very materially, and so proved to be rather a providence than a punishment.
The long-expected train of wagons arrived on the nineteenth, and there was speedily a general bustle in the camp, though never a lady near. We all set to work unloading the bales, cording and preparing them for packing, and making other necessary arrangements for prosecuting our journey. Our party now amounted to forty-five persons, and we had above a hundred beasts of burden. The men were supplied with arms, ammunition, pans, kettles, etc., and divided into six messes, each of which received its proportionate share of provisions, with an intimation that they must be carefully husbanded, as nothing more could be obtained until we reached the Council Bluffs, the intervening country being an

unpeopled waste or wilderness. Pleasant intelligence this for the stomach, and some went supperless to bed - no, blanket - for fear they would otherwise have no breakfast on some subsequent morning. At last, all was in readiness, and early the following day we were on the march. Passing the boundary of those two great states, Missouri and Misery, and leaving the forest bordering the river, we emerged into an almost limitless prairie, embroidered with woodland stripes and dots, fringing and skirting the streams and rivulets by which it was not inelegantly intersected and adorned. The day was bright and fair, and this early part of our travel might have been pleasant, but for the unceasing annoyance of our mules, who seized every opportunity, and indeed when occasion was wanting, took the responsibility of making one, to give us trouble and vexation. Some were content to display the stupidity for which their sires are so proverbial, but the greater part amused themselves with the most provoking tricks of legerdemain, such as dexterously and by some cabalistic movement, tossing their packs, (which were lashed on,) into a mud-hole, or turning them by a practised juggle from their backs to between their legs, which, having accomplished, they scampered off in high glee, or stopped and commenced kicking, floundering, pawing, and bellowing, as if they were any thing but delighted with the result of their merry humours. Job himself would have yielded to the luxury of reviling, had his patience been tried by the management of a drove of packed mules, and it may be esteemed fortunate for his reputation that Senior Nicholas had not the wit to propose such an experiment upon his even-toned temper. As the Devil is ordinarily by no means wanting in shrewdness, the omission might perhaps be set down to his credit on the score of charity, but for his abominable taste in matters of diabolical vertûe, as shown by his penchant for sanguinary signatures to all compacts and bonds for bad behavior made with or exacted by him, in the course of his "regular dealings" with mankind, and hence it must be considered a clear case of ignorance or oversight, that this test, compared to which there is toleration for boils even, was not applied. A wicked wag at my elbow, inquires with an affectation of much interest, if Satan, having in the case of the good man Job, failed so signally to keep his word, was not liable to an action on the case for a breach of promise. I of course decline answering, and refer him to those more skilled in legal casuistry for a reply. Of all bores in the world, your quizzing, carping, text-torturing sceptic is the worst - next to mule driving; and those confounded mules would bore a two inch auger hole through the meekness of Moses himself, were he their master. Such kicks, caperings, perverseness and obstinacy! the task of St. Dunstan was a play-spell to this teazing, tormenting tax upon one's time and patience. The man in the song, who "Had a donkey wot wouldn't go," and yet didn't "wallop him," was a miracle of forbearance and - but such people live only in song!
Well, in spite of the obstinacy of our mules, night came at last, and we halted on the margin of a pretty rippling stream, turned our horses loose to crop the yellow beard on the prairie face of earth, and kindled camp fires for our evening meal. O what a luxury it is to have a whole night's rest before you, after a long day of toil,

vexation, and weariness! Supper over and I indulging in reflections of a very indiscriminate kind, reposing on my elbow by the warmth of a genial blaze, when a blessless wight elbowed my repose by stumbling over me and adding an unexpected and quite too general ablution from his freshly filled kettle of water. Peace societies were not then thought of, and as I half suspected the rascal to have done it accidently by design, as an Irishman might say, I started up in order to give him, - as one good turn deserves another, - a box on the ear for his carelessness. But fear collapsed the coward's limbs, he slipped down to his knees, and my blow, just grazing the stubble of his short crop, cut the empty air and whirled me sprawling over him. There was an attitude for a philosopher! I sprang to my feet now as thoroughly enraged as I had been before drenched, but my opponent had utterly vanished, and I saw and heard nothing save the echo of a chuckle that seemed to dance on the still quivering leaves of a bush he must have brushed in his flight. However, I had my revenge for a few hours later I thrashed him soundly - in a dream!

In the morning we collected our horses and pack animals, and after breakfast continued on our journey across the prairie which we found to be lacquered with numerous trails or paths beaten by herds of buffaloes, that formerly grazed these plains, vestiges of which were still every where to be seen. One of these trails bearing to the westward we followed until it terminated in an impenetrable thicket, when our bewildered guide struck off to the northward, on a hunt, as some one facetiously remarked, after the Great Bear, which he had the good fortune to find, though not, as may be supposed, until some time after dusk. We halted for the night in a beautiful grove near a fine spring, and had the inexpressible pleasure of ascertaining that it was a capital watering place, a fact that was fully proved by the torrents that poured down like another deluge, the whole night, and prevented us from getting a single moment's sleep. Some of our people took, from this cold water movement, such a decided distaste for the pure element that they could not bear to drink a single drop, for a long time after, that is when anything better, as rum or whiskey, could be had. For my own part the surfeit did not produce nausea, and I still loved the sparkling liquid, but I must confess in more moderate abundance and from any spring rather than a spring shower.

Chapter III

We left ourself, at the close of the last chapter, in a most comfortless condition, that is to say, wet as a drowned rat, but very much consoled by the reflection that not a man in camp had a dry thread on his back. How gratifying it always is, to a person in distress, to know that his neighbours are at least as badly off as he is! There was no trouble in rousing the party that morning, for every man was up, not exactly bright to be sure, but quite early; and the number of big blazing fires, with human figures crouching and crowding round them, shifting sides and changing positions constantly, gave one no unapt conception of a certain place more than an ell in measurement, with its attendant imps and demons. Forty five persons doing duty ex necessitate rei, in the capacity of clothes-horses, had in it something indescribably ludicrous, yet, strange to say, there was not a smile on a single lip, and we all spread ourselves to dry with, the utmost imaginable gravity, specific and facial. After breakfast we gathered up our traps, literal as well as hyperbolical, and proceeded on our journey.

For several days, we met with no adventure worth relating, and though our curiosity was constantly on the stretch, to find out how it was possible for our mules to play us so many tricks as they continually did, it still remains a mystery, as much so as any other species of animal magnetism, in vogue with beings of that order. We saw herds of deer daily, now and then a herd of elk, and of deer and buffalo more bones than we cared to pick. We met also with a great variety of wild fowl, which are common to the lakes and prairies of Illinois, and to whoever can catch them besides. Innumerable small streams crossed our course, or rather we crossed them, the beds of many of which, though any thing but down, were as soft as could be desired, and much more so than suited our convenience, for they often suited us with a covering infinitely more adhesive than agreeable. Some of them we bridged over, and so passed without taking toll of their richness, but others were destitute of trees or shrubs, and because they were naked we were obliged to denude ourselves, wade over and carry both our clothes and luggage, for our horses and mules could with difficulty flounder through when eased of their loading. Of the latter it may be here observed, that however firm the bed and consequently practicable the passage of a stream might be, they invariably insisted upon not attempting to cross until relieved of their burden, and the strongest argument scarcely sufficed to overcome this repugnance to such a proceeding. "It is quite astonishing," said a weather beaten wag one day with great simplicity, "how little confidence them animals has in themselves." Singular, but our impressions were quite the contrary, and we had often occasion to remark that their organs of self-esteem and firmness must be most surprisingly developed - pro-di-gous! as Dominie Sampson would say.

On the twenty-eighth we narrowly escaped losing our horses and baggage through the carelessness of one of our men, who kindled a fire and left it notwithstanding he had been repeatedly warned of the danger of so doing. During his absence the dry grass caught the blaze, and a fresh gust in a moment fanned it to a conflagration which wrapt the whole encampment in a sheet of flame. We rushed at once to rescue the baggage, but several bales of powder and other articles were already lost to view in the devouring element that rolled and billowed over the plain. We had barely time, the flames spread with such rapidity, to seize each a bale and fly for refuge to a small sand bar, beneath a high bluff. Here we stood and gazed with agony at the curling and darting flames as they swept over the prairie, threatening destruction to our horses, in which event our situation would have been indeed deplorable. Fortunately however the wind suddenly changed, and blew with equal violence in the opposite direction, driving the mass or sheet of flames away to the eastward, and leaving us and our poor beasts free from danger.

The bales were all cased with thick cowhide and passed the fiery ordeal without injury; even our powder, though the envelopes were scorched and blackened by the blaze, escaped explosion, and we had truly reason to be thankful for our great deliverance. Two of our horses were less fortunate than their companions, for they were overtaken by the flames and completely singed, presenting an extremely ludicrous but pitiable appearance. Is it not singular that these animals, not usually wanting in sagacity or courage, should when threatened by fire so quietly submit to their fate without making a single effort to escape? A few saddles, blankets, and other articles, among which was all the extra clothing and only coat, of him whose inexcusable carelessness had thus exposed us, were lost by the fire. And this was fortunately the extent of the damage.

Resuming our journey we reached the Missouri on the thirty-first and crossed in a keel boat to Belle Vue, the trading house of Messrs. Fontenelle & Dripps, situate eight miles above the mouth of the Platte. We were here supplied with tents, which we pitched - not as the paddy did with grease - near the Papillon creek, about a mile below the fort. Our horses having become extremely weak and thin from scanty fare and hard usage, were now turned out to graze in fields of gigantic rushes which flourish in great abundance in the woodland bottoms bordering the river. As for ourselves having a long holiday before us, we employed our time in various ways, as hunting, fishing, and story telling, and making necessary preparations for continuing our route when our horses should have become sufficiently recruited to warrant them in a serviceable condition.

I shall not stop to mention all the silly things we did on the first of April, when people make such egregious fools of themselves in trying to befool others. "Oh! Ferris!" calls out one in over acted alarm, "there's a great copperhead just behind you!" "Yes, I see the rascal's face right between your two ears." Suddenly another cries in a simulated agony of terror, "Indians! Indians !" "Where ? where ?" eagerly asks some unsuspecting innocent in real fear. "April fool!" returns the wag with a chuckle, and then one tries very hard not to seem sheepish, but to look a whole folio of dignified philosophical indifference, in both of which he utterly fails as a matter of course, while the other builds a couple of triumphal arches with his eye brows, and hieroglyphs his face over with tokens of self gratulation at his successful foray, - fooled each to the top of his bent. In puerilities like these passed the day, as all-fools day usually passes, in country, camp, or court the world over. Vive la bagatelle! - hurra for nothing!

The four weeks of our stay at this point were undiversified by any occurrences worth relating, and we soon became heartily weary of the dull monotony of its daily routine, and as anxious to resume the line of march, as we had been before to hail a pause in its progression. The days dragged on heavily and slowly until the last of April came, when after packing up with the alacrity of pleasure, we packed off in high spirits and ascending a hill in rear of the trading house, bade a long but unreluctant adieu to the scene of a wasted month, glad to find our feet again in the stirrups, and our faces once more, westward ho! We soon lost sight of Belle Vue, though belle vue was ever in sight, in whatsoever direction our eyes were turned. But the same cause that rendered the prospect beautiful, namely, several recent showers, had also made the roads almost impassable. Our mules were become more mercurial than ever and played off their old pranks with a skill greatly heightened by experience, much to the annoyance and vexation of the poor Jobs, who were compelled to manage, and yet - incredible hardship! - not permitted to kill them. Here or there might be seen at almost any moment, some poor devil smeared or bespattered with mire and water until he scarcely knew himself only by report, holding on to a restive mule with one hand, and with the other endeavouring to fish out of the mud a discharged cargo, left without leave by the gallows jade whose business it was to bear the burden. These knights of the cross (the poor mule drivers) as their crosses and losses of luck and temper occasioned them to be called, were cross from morning till night and yet I doubt if they were not naturally the best natured fellows in the world; but mule driving is the d__l and there is no more to be said about it, except that I pitied them until it came my turn to share their fate, and then I pitied the tiger for his tameness. We slept that night at a fine spring ten miles north of Belle Vue, and, oh strange inconstancy of man's mood! wished ourselves back by the quiet margin of the peaceful Papillon, whose rushy border we had rushed away from but a few hours before.

"Green grow the rushes O!
Green grow "
Good night!

Chapter IV

And this is May-day, the festival of girlhood and happy youth, in many a town of many a land, where joyous hearts exulting hail its beautiful dawn, and the hours are winged and rosy with the exciting and rapturous scenes of a floral coronation. Ah, how sweetly rise in my memory the visions of fetes like these! I can almost fancy that I see one now - that again a laughing gay spirited boy I mingle in the mimic pageant, and assist at the pleasing ceremonial. There stands the rural throne, with its velvet dias, its mossy seat, and its canopy of flower-woven evergreens; there too, is the fairy-like Queen, a tall, graceful girl, the flaxen locks of whose infancy have been curled into golden ringlets, that cluster round her beautiful face, and fall in fleecy masses on her ivory shoulders, by the warm suns of some thirteen summers; and there, too, is a gallant gathering about her of maids of honour, pages, pursuivants, and - pshaw! what a fool I am to dream of scenes and seasons like those, in this far wilderness, and with these companions! Imagination! and thou, too, Memory! be silent, and weave no more the bright texture of romance!

Resuming our march, we followed a zig-zag trail through hills, and bluffs, covered with dwarf trees, and thick underbrush, for six miles, and descending into a pleasant vale, came upon the Trading-house of Mr. Cabina, eight miles below the Council Bluffs. Here we received supplies of ammunition and a "Code of Laws," with penalties annexed, for the preservation of harmony and safety, in our passage through the immense plains - that still intervene between us and the end of our journey - which are roamed and infested by hordes of savages, among whom theft and robbery are accounted any thing but crime, and whose scruples on the score of murder are scarcely a sufficient shield against the knife or the tomahawk. Strength and courage alone, command their respect - they have no sympathy for trust, no pity for weakness. By the strong hand they live, and by the strong hand only are they awed. Our traveling code of "pains and penalties" was signed by Mr. Fontenelle, a veteran leader in the mountain service, who now assumed the direction of affairs and in all things showed himself to be an experienced, able, and efficient commander.

After a brief interval of rest, refreshment, and preparation, the word was given to march, and, leaving Mr. Cabina, his trading house, and the Missouri, we struck off across the prairie until evening, when we pitched our camp on the Papillon, twenty miles above its mouth. Next day we reached a branch of the Loup Fork, called the Elk-horn River - a clear, deep, rapid stream, fifty paces in width - and constructed a boat-frame of willow, which we covered with dressed buffalo-skins, sewn together for the purpose. After some trouble in adjusting and securing the parts, our boat was finished, and launched, but unfortunately the skins proved to have been spoiled and soon came to pieces. We had but one resource left, and that to ford the river, which was effected at a point where the greatest depth did not exceed four feet. Stripping ourselves, and wading back and forth we transported our baggage on our backs, piece-meal, whilst our horses were forced to swim over at another place. The water was quite chill, and as if to make the toil of crossing doubly unpleasant, we were showered with a storm of sleet, which belaboured our naked shoulders most unmercifully. However, we got every thing at last safely over, and as evening overtook us here, passed the night on the margin of the river. We started as usual, early on the following day, but proceeded only a few miles, when we were compelled to halt at a place called "The Hole," in consequence of a severe storm of sleet, accompanied by a fierce northern gale, which continued with unabated fury till the morning of the fifth. We began to grow familiar with hardships, as may well be imagined, from the toil, danger, and exposure, of scenes like these, but such weather was still - awful unpleasant!

The country now presented a boundless gently-rolling prairie, in one complete mantle of green, laced with occasional dark stripes of woodland, that border and outline the mazy courses of rivulets, which flow from every dell and hollow. Wild onions abound on the margin of all these streams, as the lovers of that valuable and very fragrant esculent may be pleased to learn; but I botanized no further. On the fifth we continued our march, with the bright sun of a beautiful day smiling upon and encouraging our journey.

Up to this period, we encamped without order, helter-skelter, just as it happened, allowing our horses to run loose night and day; but now, when we halted for the night, our camp assumed a somewhat martial appearance. The order of its arrangement was this, - a space of fifty yards square was marked out, one side of which was always along the brink of some stream. Four of our tents occupied the corners, and of the remaining four, one was placed in the middle of each side. The intervening spaces between the tents were barricaded by a breast-work formed of our baggage and horse furniture. The space within the square, was dotted with the iron heads of nearly two hundred hard wood pins, each one foot in length, and one and three-fourths inches in diameter, drove into the ground, to which our horses and mules were fastened. Each man was provided with a wooden mallet to drive the pins with, and when, just before sunset, all were put into requisition, such a din as they created, would be a caution to Paganini. Immediately after sundown, the words "catch up," resounded through camp, all hands flew to the horses, and all was noise and bustle for some minutes. Forty odd of us 'cordelling' our stubborn mules, - who the more you want them to go, the more they won't - into camp, with oaths and curses, not only loud, but deep - it was wicked, but, poor fellows they couldn't help it! - might have been seen, if one could for laughter have kept his eyes open, upon any such occasion. A few moments and all was quiet again, horses and mules securely fastened to their respective pickets, and the men at their tents, seated around kettles of boiled pork and corn, with pans, spoons, and grinders in motion. Keen hunger made us relish the repast, which else the very dogs had refused, - however all contented themselves as well as they might with such fare, looking forward with a sort of dreamy delight to the time when rich heaps of fat buffalo meat, should grace and garnish our encampments.

After supper we reclined on our elbows about the fire, produced our pipes and tuned them to a smoke, recounted tales, puffed ourselves, and old times, and quizzed, joked and jested with one another until eight o'clock, when our humour was interrupted by the cry "turn out the first guard, " whereupon six of our companions, jumped up, seized their guns and blankets, and presently commenced strutting around camp, rifle in hand, while the rest retired not only to sleep, but also to be awakened, in the midst perhaps, of a pleasing dream, by a rough shake of the shoulder, and those most detestable words, "get up, sir, it is your watch, " - and capital time those watches keep too, except that they are apt to run a little too fast. Two hours, two mortal long hours, wrapped in your blanket may you sit on the prairie without fire, but with your rifle across your knees, and watch the stars, the moon, the clouds, or the waving grass, not forgetting to answer the watch-word repeated every half hour, by six poor wretches like yourself, "all's well." Rain or shine, wet to the skin or not, half starved with cold or hunger, no matter what, still you hear and echo those most applicable words, with perhaps, as once in a woeful storm of sleet, the rhyming jingling comment of some uneasy sleeper, - " 'Tis false as h__l," the truth of which in your heart you are forced to admit. At the expiration of two hours another takes your place, and you may crawl to rest, to be brought again to your feet at day light by the cry "léve, léve," (get up). Three or four of the morning guard are ordered at dawn to scour the neighbouring hills on horseback, when if they discover nothing unusual, the horses are turned out to graze, under the charge of the "horse day guard," and the rest of the party cluster round their camp-fires to smoke or watch the bubbling kettle, till the morning meal. After breakfast all are busily employed in folding up their tents, pulling down the breast-works, and arranging the luggage so as to require as little time as possible for "loading up." When the sun is something over an hour high, the order "catch up," is again heard, and all hasten to catch and tow their animals into camp. Patience and forbearance, if you are blessed with those amiable qualities, will now be tested to the uttermost, supposing you to be honoured with the charge of two or more of those mongrel brutes with shrill voices and long ears. Few exist but will strive to do you an injury by some infernal cantrap or other. One bites your leg while you fasten the saddle girth, another kicks you while you arrange the croupper, a third stands quietly until his lading is nearly completed, and then suddenly starts and flounces until he throws every thing off, a fourth at the same interesting point stamps upon your foot, breaks away, and scampers off into the prairie, strewing the way with his burden, a fifth refuses to be loaded at all, and a sixth to stand still, be led or driven. In short there is no end to their tricks and caperings. But I spare the recital. Any one of the party having completed his arrangements for departure assists his messmates, and in half an hour or so, all are ready for marching orders, when our leaders take the front, and proceed at a fast walk, while we fall into line and follow, leading our pack horses, and carrying our guns before us across the saddle. At noon we halt for a couple of hours, after which we journey on until the sun appears but an hour and a half or such a matter above the horizon, when we stop for the night, turn out our horses, after "hobbling" them, by tying the fore legs together to prevent them running away in case of an alarm, and arrange and fortify our encampment, as above related.

Chapter V

We saw on the seventeenth several prong-horned antelopes; a timid, fleet, and beautiful animal, peculiar I believe to the region of the Rocky Mountains. Much I had heard and read of the swiftness and graceful motion of the antelope, but had no conception of the exquisite ease of its airy, floating perfection of movement, until I saw these glide away with the light and sylphic step of the down-footed zephyr, that scarcely touches the lawn over which it trips so sweetly and so swift. I can now understand, what I never could realize before, the poetry of motion. We reached on the following day a wide shallow stream called the Loup Fork, which rises near the Black hills, and flows eastward about five hundred miles, parallel with the Platte, into which it empties forty or fifty miles above the Missouri. We found no little difficulty in fording it, in consequence of the quick sands of which its bed is composed, giving way so readily beneath the pressure of our feet. At noon, however, all were safely across, and for the rest of the day we skirted along its southern margin. The following afternoon we passed a Pawnee village situated on the opposite bank of the river, and sent, as customary, a present of tobacco, powder, balls etc., to these tribute-taking lords of forest, field and flood, the heart of whose wild dominion we are now traversing. In the evening the principal chief a fine looking, hardy, and certainly hearty old codger, and two of his people, came with our messenger to pay us a visit and acknowledge our courtesy, when the pipe of peace was smoked with all becoming gravity, and he was so well pleased with his reception and our hospitality that he passed the night with us. The same evening one of our men by the name of Perkins, was severely burned by the accidental explosion of his powder horn. On the next day we reached and crossed the Platte river, which is here nearly a mile wide, but so shallow as to be fordable. It is full of low sleepy islands, and bounded on either side by rich bottom lands, often a mile two in breadth, but little higher than the stream itself, and apparently quite as level. The bed of this river is also formed of quicksands which are always shifting, and give its waters that muddy consistence so remarkable in the Missouri. Beyond the bottoms a rolling sandy prairie stretches its lazy level, but scantily covered with a coarse short grass, and even now and then in barren spots as nude as an antique statue destitute of the seemliness of a fig leaf. Occasional groves of aspen and cotton-wood deck the islands and bottoms of the Platte, and these are the only varieties of timber to be found.

Scarcely had we got under way on the morning of the eleventh, when we discovered several mounted Indians approaching at full speed, who soon gave us to understand that a large party of their people were close at hand coming to trade with us. Mr. Fontenelle not doubting but that they came for the express purpose of plundering us, immediately ordered a halt, and made preparations to give them a reception more warm than welcome. We picketed and hobbled our horses, examined our guns, and were directed to be ready for the worst. Hardly were these hasty preliminaries arranged, when the Indians, a large body of well mounted fine ferocious looking fellows, dashed in sight at the top of their speed. We formed a line in front of our baggage, all wide awake for a nice cosy little game of ball, and quietly waited their approach. Our suspense was not of long duration for they whirled up in a breath to speaking distance and were ordered to stand, which they did in mid career, throwing their horses back upon their haunches, and halting about two hundred yards in advance of us, when their chief commenced a loud harrangue in the choicest guttural that could be conceived, much to our edification and delight. They appeared to be about one hundred and fifty strong, (only thrice our number), all admirably mounted, and all armed with bows and arrows, and spears, and a few with guns. They wore buffalo robes about their middle, but from the waist upwards were all magnificently naked. A few had on leggins of dressed skins, but generally save their robes and moccasins, they were just as nature made them, except in the

matter of grease and paint. After some introductory chatterings, they informed us that they were on a hunting expedition for buffalo, that they intended us no harm, but on the contrary wished to trade with us in amity. They were then permitted to come up, and exchange a few skins, moccasins, etc. for knives, vermillion, and tobacco, pilfering the while every thing they could lay their hands upon without being discovered. The reciprocity of this kind of commerce being as the Paddy said, all on one side, we soon got tired of it, and unceremoniously packed up and off, and left them gazing after us in no small astonishment.
On the fourteenth, hurrah, boys! we saw a buffalo; a solitary, stately old chap, who did not wait an invitation to dinner, but toddled off with his tail in the air. We saw on the sixteenth a small herd of ten or twelve, and had the luck to kill one of them. It was a patriarchal fellow, poor and tough, but what of that? we had a roast presently, and champed the gristle with a zest. Hunger is said to be a capital sauce, and if so our meal was well seasoned, for we had been living for some days on boiled corn alone, and had the grace to thank heaven for meat of any quality. Our hunters killed also several antelopes, but they were equally poor, and on the whole we rather preferred the balance of the buffalo for supper. People soon learn to be dainty, when they have a choice of viands. Next day, oh, there they were, thousands and thousands of them! Far as the eye could reach the prairie was literally covered, and not only covered but crowded with them. In very sooth it was a gallant show; a vast expanse of moving, plunging, rolling, rushing life - a literal sea of dark forms, with still pools, sweeping currents, and heaving billows, and all the grades of movement from calm repose to wild agitation. The air was filled with dust and bellowings, the prairie was alive with animation, - I never realized before the majesty and power of the mighty tides of life that heave and surge in all great gatherings of human or brute creation. The scene had here a wild sublimity of aspect, that charmed the eye with a spell of power, while the natural sympathy of life with life made the pulse bound and almost madden with excitement. Jove but it was glorious! and the next day too, the dense masses pressed on in such vast numbers, that we were compelled to halt, and let them pass to avoid being overrun by them in a literal sense. On the following day also, the number seemed if possible more countless than before, surpassing even the prairie- blackening accounts of those who had been here before us, and whose strange tales it had been our wont to believe the natural extravagance of a mere travellers' turn for romancing, but they must have been true, for such a scene as this our language wants words to describe, much less to exaggerate. On, on, still on, the black masses come and thicken - an ebless deluge of life is moving and swelling around us!

Chapter VI

Since leaving the Loup Fork we have seen very little timber, and latterly none at all. We have, however, hitherto found plenty of drift-wood along the banks of the river, but to-day, the nineteenth, there is not a stick of any description to be seen, and as the only resource, we are compelled to use as a substitute for fuel, the dried excrement of buffalo, of which, fortunately, the prairie furnishes an abundant supply. I do not, by any means, take it upon myself to defend the position, but certainly some of the veterans of the party affirm that our cooking exhibits a decided improvement, which they attribute to this cause, and to no other. That our steaks are particularly savoury I can bear witness.

At our noon encampment on the twenty-first, we discovered several objects on the brow of a neighboring bluff, which at first we took to be antelopes, but were soon undeceived, for they speedily transformed and multiplied themselves into several hundreds of Indians, who came rushing like a torrent down upon us. All was now excitement and confusion. We hastily collected our cattle, drove them into camp, and fastened them, built a breastwork of our baggage, primed our guns afresh, and prepared to stand upon our defence. The Indians by this time came up, made signs of friendship, and gave us to understand that they were Sioux. They formed a semicircle in front of our position, and displayed four American flags. Many of them had on long scarlet coats, trimmed with gold and silver lace, leggins and mocasins richly, though fantastically ornamented, and gay caps of feathers. Some wore painted buffalo robes, and all presented a lively, dashing appearance. They were, without exception, all finely mounted; and all armed - some with swords, shields, and lances, others with bows and arrows, and a few with guns.

After some consultation among themselves, they informed us, with much gravity, that it was customary for whites passing through their country to propitiate their friendship by a small present, which was immediately acceded to, and a liberal gift of ammunition, knives, trinkets, and paints bestowed. Several of their chiefs passed through our camp while this was doing, and we observed that some of them wore large silver medals. During the whole time the interview lasted, the rain came down in torrents, and the air was besides extremely cold. Wet to the skin, and chilled to the very marrow, we were compelled to stand to our posts, with limbs shivering, and teeth chattering, while the Indians warmed themselves at fires made of the buffalo dung we had collected. I never in my life had a stronger desire to pull trigger on a red skin than now, but they gave us no sufficient provocation to authorise hostilities; and to our great relief, after getting from us all they could beg, and stealing all they could slyly lay their hands on, they took their departure, and returned to their own camp.

The following day was raw, wet, and cold, and the "prairie chips" having now become so saturated with water that they could not be coaxed to burn, we had no alternative but to freeze or move camp. Preferring the latter, we resumed our weary march, and fortunately, after six miles travel, found a welcome plenty of drift-wood, when we again halted to enjoy the luxury of a good fire in a rain storm in the open prairie. Blessings on thy head, O Prometheus! that we have even the one comfort of a cheerful blaze.

We saw a wild horse next day, on the opposite side of the river, and made an effort to catch him, but did not succeed. An Indian, ordinarily well mounted, would have caught him with a noose almost in no time; but luckily for him, we were not Indians. One singular fact, often remarked, but never, that I know of, chronicled, is this, that a horse carrying a rider will easily overtake one not mounted, though naturally much the fleetest. I cannot account for this, but it is nevertheless true, and can be proved by an abundance of testimony.

We traversed on the twenty-fourth, a narrow tract of country, covered with light sand, and destitute of every kind of vegetation, save a species of strong grass, covered with knot-like protuberances, which were armed with sharp thorns that pierce the foot through the best of moccasins. These grass-knots are called "Sand-burrs," and were a source of great inconvenience to several poor fellows who, as a punishment for having slept on guard, were compelled to trudge along on foot behind the cavalcade.

On the twenty-fifth we saw a herd of wild horses, which however, did not wait a very near approach, but dashed off, and were soon lost in the distance. We had a visit in the afternoon from three Sioux, who came into camp, and reported that a large collection of Arrapahoes and Gros Ventres lay in wait for us at the Black Hills, determined to give battle to all parties of whites who should attempt to pass them. It was little uneasiness this intelligence gave to the men of our party; we were growing wolfish after some kind of excitement, and would have fought a whole raft of them in our then present humour, for the recreation of a play spell. It may well be questioned, however, if our leaders, who had the responsibility of a double charge, were quite so indifferent to the matter. But n'importe.

We reached on the following day the "Nose Mountain," or as it is more commonly called, the "Chimney," a singular mound, which has the form of an inverted funnel, is half a mile in circumference at the base, and rises to the height of three hundred feet. It is situated on the southern margin of the North Fork of the Platte, in the vicinity of several high bluffs, to which it was evidently once attached; is on all sides inaccessible, and appears at the distance of fifty miles shooting up from the prairie in solitary grandeur, like the limbless trunk of a gigantic tree. It is five hundred miles west from the Council Bluffs.

We encamped on the twenty-seventh opposite to "Scott's Bluffs," so called in respect to the memory of a young man who was left here alone to die a few years previous. He was a clerk in a company returning from the mountains, the leader of which found it necessary to leave him behind at a place some distance above this point, in consequence of a severe illness which rendered him unable to ride. He was consequently placed in a bullhide boat, in charge of two men, who had orders to convey him by water down to these bluffs, where the leader of the party promised to await their coming. After a weary and hazardous voyage, they reached the appointed rendezvous, and found to their surprise and bitter disappointment, that the company had continued on down the river without stopping for them to overtake and join it.

Left thus in the heart of a wild wilderness, hundreds of miles from any point where assistance or succour could be obtained, and surrounded by predatory bands of savages thirsting for blood and plunder, could any condition be deemed more hopeless or deplorable? They had, moreover, in descending the river, met with some accident, either the loss of their arms or powder, by the upsetting of their boat, which deprived them of the means of procuring subsistence or defending their lives in case of discovery and attack. This unhappy circumstance, added to the fact that the river was filled with innumerable shoals and sand-bars, by which its navigation was rendered almost impracticable, determined them to forsake their charge and boat together, and push on night and day until they should overtake the company, which they did on the second or third day afterward.

The reason given by the leader of the company for not fulfilling his promise, was that his men were starving, no game could be found, and he was compelled to proceed in quest of buffalo. Poor Scott! We will not attempt to picture what his thoughts must have been after this cruel abandonment, nor harrow up the feelings of the reader, by a recital of what agonies he must have suffered before death put an end to his misery.

The bones of a human being were found the spring following, on the opposite side of the river, which were supposed to be the remains of Scott. It was conjectured that in the energy of a dying despair, he had found strength to carry him across the stream, and then had staggered about the prairie, till God in pity took him to himself.

Such are among the sad chances to which the life of the Rocky Mountain adventurer is exposed.

Chapter VII

At about noon on the twenty-eighth we discovered a village of Indians, on the south side of the river five miles above, and sent three men forward to watch their movements whilst we made the necessary preparations for defence. In a short time our spies returned, closely following by about fifty of the Indians, who dashed up in a cloud, and gave us to understand that they were "Chayennes." They repeated the story told by the Sioux, respecting the Arrappahoes Gros Ventres, and remaining about us till night when all but one disappeared.

In the course of the evening it was whispered about that the Indian in camp was an Arrappahoe, (with whom we were at war,) and one of the men became so excited on the subject that he requested permission to shoot him, but was of course refused.

During the night this individual, with two others, made an attempt to desert, but was detained by the guard. To such a pitch of desperation were his feelings wrought up that on the following morning he left us to return alone to St. Louis, notwithstanding, as he acknowledged, fear alone had impelled him to attempt desertion. It was a singular case, the very excess of cowardice having determined him to an undertaking from which the boldest would have shrunk appalled.

We afterwards heard that he succeeded in reaching St. Louis alive, but that he suffered the extreme of misery both from starvation and maltreatment of the Indians, some of whom seized him near the Council Bluffs, stripped him entirely naked, scourged him most unmercifully, and then let him go. In this situation he found his way to the garrison near the Platte, more dead than alive. Here he was kindly received, supplied with food and clothing, and nursed up until his health was quite recruited, when he returned to St. Louis, and reported that the company had been attacked and defeated by the Indians, himself alone escaping. On the day he left us we reached a fine grove of cotton wood trees of which we made a horse pen - this is always done in the Indian country when timber can be obtained, as a necessary protection for our cattle, in case of attack. Save a few isolated trees, this is the only timber we have seen for fifteen days.

We discovered on the thirtieth, a solitary Indian lodge, pitched in a grove of aspen trees, which, as it was the first I had seen, was an object of some curiosity. The manner of its construction was this: - thirteen straight pine poles were placed equidistant from each other in the circumference of a circle, ten or twelve feet in diameter, and made to meet in a point eleven feet from the ground, where four, crossing a foot from the end, are tied together, to support the rest. The conical frame thus formed is covered with dressed buffalo skins, cut and sewed together in a proper shape, which much resembles the shape of a coat. A pole fastened at top and bottom to this covering serves to raise it by, the top of which is allowed to rest against the others. Then the loose sides are drawn around the frame and fastened together with strings or wooden pins, to the height of seven feet, except that an oval aperture three feet high is left for an entrance. Above the closed parts, are two projecting wings or corners with pockets on the outside for the reception of two poles, calculated to piece them in various positions, in order to avoid the smoke, which but for some such contrivance, would greatly incommode the inmates, particularly if the wind should happen to come from an unfavourable quarter. The bottom of the covering is then secured to the ground on the outside with wooden pins, and the lodge is thus complete. If it be well pitched the covering sets smoothly to the poles and is tight as a drum head. A skin fixed to hang loosely over the aperture serves the purpose of a door, and this concludes the description of any lodge hereafter mentioned, though some are larger and others less in proportion.

As we approached the lodge the first object that presented itself was the lifeless body of a male child about four years old. It was lying on the ground a few paces from the lodge, and was horribly maimed and disfigured, evidently by repeated blows with a club, it bore also the mark of a deep wide stab in the left side. Within the lodge on a raised platform lay the scalpless bodies of two grown Indians, with their instruments of war and the chase beside them - it being the Indian custom to bury with the dead such articles as they believe will be required on a journey to the land of Spirits. Both the bodies were hacked and mangled in a manner truly savage and revolting. They were Chayennes, and had been killed in a battle with the Crows, five days previous. The child was a prisoner taken from the Crows the preceding winter, and was thus barbarously murdered by way of retaliation. Achilles sacrificing at the tomb of Patroclus - is both a precedent and a parallel. Poetry has almost hallowed the cruelty of the Greek, but the inhumanity of the savage is still fearfully conspicuous; yet which was the worst, the refined Hellenian or the barbarous Chayenne? We crossed the Platte in bull-hide canoes, on the second of June, and encamped a short distance above the mouth of Laramie's Fork, at the foot of the Black Hills, six hundred miles west of the Council Bluffs. Laramie's Fork rises in the Black Hills, between the northern and southern forks of the Platte, and falls into the former after a northeast course of six hundred miles. The rich bottoms bordering this stream are decked with dense groves of slender aspen, and occasional tall and stately cotton woods.

Since passing the Sioux country we have seen herds of buffalo almost daily, but never in such countless numbers as then astonished our sight. Our hunters kill more or less of them every day, and they form the staple article of food; but they are still poor and tough, and would hardly be considered eatable could any thing else be procured, which is not the case.

The Black Hills are a chain of mountains less remarkable for height than for scenery, which is of the most romantic order. They extend up the Platte one hundred miles, and are noted as a place of refuge and concealment for marauding Indians, - they form consequently a dangerous pass for hunting and trading parties. They are partially covered with pine and cedar shrubbery, which gives them when viewed from a distance, a dark forbidding appearance, and hence their name. On a nearer approach, however, they present a less repulsive aspect, and finally exhibit a pleasing variety of shapes, and colours, slopes and dells, bluffs and ravines, which together form occasional landscapes of singular picturesqueness and beauty. Some of these hills are composed of a deep crimson-coloured sand stone, others of a bright yellow, grey, white or brown rocky formation, and all partially covered with soil. Some are as bald of vegetation as the naked prairie, and one, rearing its barren peak far above the rest, is still crowned with a diadem of snow. Entering the region of this range of hills, the Platte was seen to the right of our trail, winding its devious way, at times through fine timbered bottoms, again between dark walls of cut rock, and occasionally through beautiful unwooded valleys occupied by herds of buffalo quietly grazing and all unconscious of the approach of death in the form and guise of old Sonsosay, our veteran hunter, who might have been seen crawling like a snake through the long grass, until a sudden burst of thunder starting herds and echoes from their repose, showed that he had them within reach of his unerring rifle, where horns and hoofs were alike unavailing.

Chapter VIII

On the eighth, we saw for the first time, a grizzly bear, a large fierce formidable animal, the most sagacious, most powerful, and most to be feared of all the North

American quadrupeds. We shall have occasion elsewhere to note instances of the prowess, cunning and courage of this remarkable animal, and shall relate, in their proper connexion, some of the many anecdotes concerning it, which are current among the Indians and trappers of the Rocky Mountains, the stock of which is constantly increasing, as adventure goes on, and brute and human meet in mutual strife. The one we saw, was at a distance, and looked nearly as large as a buffalo, for which they are often mistaken, even by experienced hunters.

We re-crossed the Platte again, on the eleventh, at the Red Hills, - these are two high cherry-red points of rock, separated by the river, which here turns away to the southward. On the following day we left the Platte and the Black Hills together, and pursued our never-varying course, westward, through a sandy plain, covered with wild sage, and at evening encamped near a fine spring. Next day's march brought us to Sweet-Water River, which rises in the southeastern extremity of the Wind Mountains, and flows eastward one hundred and fifty miles, falling into the Platte, a few miles above the Red Hills. This river owes its name to the accidental drowning in it of a mule loaded with sugar, some years since. We halted at evening under the lee of an immense rock half imbedded in the earth, which is nearly a mile in circumference, and from one to two hundred feet in height. It bears the name of Rock Independence, from the circumstance of a party having several years ago passed a fourth of July, with appropriate festivities, under its ample shade. Except a chain of rocks, some of which closely resemble hay-stacks, running parallel with Sweet Water, the face of the country is a barren, sandy, rolling prairie, destitute of trees and bushes, and every species of vegetation, save occasional patches of coarse grass and wild sage, and scattering clusters of dwarf willows, on the margin of the river.

On the fourteenth, we passed a small lake, highly impregnated with glauber salts, the efflorescence of which, covers the margin of the lake to the depth of several inches, and appears at a distance like snow. We made a cache on the nineteenth, of some goods, intended for future trading with Crow Indians, who rove at some seasons, on the tract of country we are now passing. Cache, derived from the French verb cacher, to conceal, is applied in this region to an excavation for the reception of goods or furs, commonly made in the following manner. A proper place being selected, which is usually near the border of some stream, where the bank is high enough to be in no danger of inundation, a round hole two feet in diameter is carried down to a depth of three feet, when it is gradually enlarged, and deepened until it becomes sufficiently capacious to contain whatever is destined to be stored in it. The bottom is then covered with sticks to prevent the bales from touching the ground, as otherwise they would soon contract moisture, become mouldy, and rot. The same precautions are observed to preserve them intact from the walls of the cave. When all is snugly deposited and stowed in, valueless skins are spread over the top, for the same excellent purpose, and the mouth is then closed up with earth and stones, beat down as hard as possible, to hinder it from settling or sinking in. The surplus earth taken out, is carefully gathered up and thrown into the stream, and the cache finally completed, by replacing stones and tufts of grass, so as to present the same uniform appearance, as the surrounding surface. If the cache is made in a hard clay blu. and the goods perfectly dry when put in, they will keep years without damage. At this period we were in view of the Wind Mountains, which were seen stretching away to the northward, their bleak summits mantled over with a heavy covering of snow.

On the twentieth, we reached a fountain source of the Sweet Water, near a high, square, table-like mound, called the Pilot Butte; and the next day ascended an irregular plain, in which streams have their rise, that flow into both the Atlantic and Pacific Oceans, halting at night on the Sandy, a small river that takes its name from the barren country through which it runs. It has its source in the south-eastern point of the Wind Mountains, where also the Sweet Water, Platte, and Wind River of the Bighorn, take their rise, and empties into Green River after a south west course of sixty miles. From the dividing plain or ridge, we saw vast chains of snow-crowned mountains, stretching far away to the west and northward, and revealing but too plainly the toil and hazard that await our future progress. The southern point of the Wind Mountains, rose bluffly to the northeast, distant fifteen miles, and, strangely contrasting their snowy summits with the dark forests of pines that line and encircle their base, were seen stretching away to the northwestward the looming shapes of this range or spur of the far-reaching Andes, until their dark forms and dazzling crests were lost in the distance, blending in the haze and mingling with the clouds. After a weary march, on the twenty-first, we reached Green River, a fine, clear, deep and rapid stream, one hundred and fifty yards wide, which takes its rise in the Wind Mountains, with the sources of Lewis River and the Yellow Stone, and flows south-east, south, and finally south-west, four hundred miles, to its junction with Grand River, when it becomes the Rio Colorado of the West, one of the most magnificent streams in the world, and descending the mountains, rolls its sublime volume away, many hundreds of miles through Upper and Lower California, until at last it reaches and empties into the gulf of that name. From the southern point of the Wind Mountains, one or two snowy peaks rise, dimly visible, far to the southward: - within the intervening space, a broken, sandy plain, perfectly practicable for loaded wagons, which may cross it without the least obstruction, - separates the northern waters of the Platte, from those of the Colorado.

We crossed Green River on the twenty-sixth in bull-hide canoes, and halted for the night on its western margin, where we were nearly victimized by moschetoes, which during the five days of our vicinity to this stream, kept sucking at the vital currents in our veins in spite of every precaution that could be taken. Leaving Green River the next day, we encamped after a hard journey of twenty-five miles, on one of its branches, called Ham's Fork. From this point, several persons were despatched in different directions in quest of a party of hunters and trappers, called Free Men, from the circumstances of their not being connected with either of the rival Fur Companies, but holding themselves at liberty to trade with one or all. They rove through this savage and desolate as the mountain air, leading a venturous and dangerous life, governed save their own wild impulses, and bounding their desires and wishes ir own good rifles and traps may serve them to procure. Strange, that ..... find so strong and fascinating a charm in this rude nomadic, and hazardous mode of life, as to estrange themselves from home, country, friends, and all the comforts, elegances, and privileges of civilization; but so it is, the toil, the danger, the loneliness, the deprivation of this condition of being, fraught with all its disadvantages, and replete with peril, is, they think, more than compensated by the lawless freedom, and the stirring excitement, incident to their situation and pursuits. The very danger has its attraction, and the courage and cunning, and skill, and watchfulness made necessary by the difficulties they have to overcome, the privations they are forced to contend with, and the perils against which they must guard, become at once their pride and boast. A strange, wild, terrible, romantic, hard, and exciting life they lead, with alternate plenty and starvation, activity and repose, safety and alarm, and all the other adjuncts that belong to so vagrant a condition, in a harsh, barren, untamed, and fearful region of desert, plain, and mountain. Yet so attached to it do they become, that few ever leave it, and they deem themselves, nay are, with all these bars against them, far happier than the in-dwellers of towns and cities, with all the gay and giddy whirl of fashion's mad delusions in their train.

Continuing our journey, we passed up Ham's Fork thirty miles, and then made a halt until all our people returned, who reported that no traces of the Free Men could be found. We then resumed our march, and on the seventh of July, ascended a steep snow-clad pine-covered mountain, when we came in view of a beautiful valley, watered by a shining serpentine river, and grazed by tranquil herds of buffalo. At evening we halted on the margin of Bear River, after a very fatiguing and toilsome march of thirty miles. This river is from fifty to eighty yards in breadth, clear and deep, with a gentle current, and is bordered by fertile though woodless bottoms. It rises in the Eut Mountains, and flows northward above one hundred miles, when it turns to the westward, and after a further course of seventy-five miles, discharges itself into the Big Lake.

We killed here a great many buffalo, which were all in good condition, and feasted, as may be supposed, luxuriously upon the delicate tongues, rich humps, fat roasts, and savoury steaks of this noble and excellent species of game. Heretofore we had found the meat of the poor buffalo the worst diet imaginable, and in fact grew meagre and gaunt in the midst of plenty and profusion. But in proportion as they became fat, we grew strong and hearty, and now not one of us but is ready to insist that no other kind of meat can compare with that of the female bison, in good condition. With it we require no seasoning; we boil, roast, or fry it, as we please, and live upon it solely, without bread or vegetables of any kind, and what seems most singular, we never tire of or disrelish it, which would be the case with almost any other meat, after living upon it exclusively for a few days. Perhaps the reason why the flesh of buffalo is so superior to the beef of the United States, may be found in the fact, that during the severities of winter, they become reduced to mere skeletons, and thrive with the grass in spring, mending up constantly as the season advances, until in summer, their bones are thickly enveloped with an entire new coat of flesh and fat.

While we remained in the neighbourhood of Green River, we were again exceedingly annoyed by moschetoes. They appeared in clouds both in the morning and evening, but disappeared in the heat of the day, and with the sun at night. Parties were here a second time sent out in various directions, in search of the Free Men, but they all returned again unsuccessful. Some of them saw an encampment of a party of Indians, who had passed two days before, about Sixty miles above on this stream. They were supposed to be about one hundred and fifty strong, and were evidently on some expedition that required great secrecy and caution. They encamped in a very small circle, and removed every thing from camp, that would lead to a discovery of their nation. When they departed, they went into the hills, and were so cautious, that our spies found it not only impossible to follow the trail, but even to designate the course they had taken. The trails and encampments of a party of hunters who had passed very early in the spring, were also seen. Nothing else unusual was observed.

Chapter IX

On the sixteenth, I departed with Mr. Dripps and three others for Cache Valley. We passed up the river a few miles, crossed, and followed a rivulet westward to its source in the mountain, which we then ascended to its summit. The crest of the mountain was ornamented with a few scattering cedars, here and there a small grove of aspen, and occasional patches of wild sage. From this elevation bleak snow-clad pyramidic peaks of granite were beheld in all directions jutting into the clouds. Stern, solemn, majestic, rose on every side these giant forms, overlooking and guarding the army of lesser hills and mountains that lay encamped below, and pointing proudly up their snow-sheeted crests, on which the stars at evening light the sentinel fires of ages.

From the precipitous western side of the height on which we stood, one of the most agreeable prospects imaginable, saluted and blessed our vision. It was the Little Lake, which from the foot of the mountain beneath us, stretches away to the northward washing the base of the cordillera that invests it. It is fifteen miles long and about eight in breadth and like Nemi,

"Navelled in the hills,"

 for it is entirely surrounded by lofty mountains, of which those on the western side are crowned with eternal snow. It gathers its waters from hundreds of rivulets that come dancing and flashing down the mountains, and streams that issue not unfrequently from subterranean fountains beneath them. At the head of the lake opposite, and below us, lay a delightful valley of several miles extent, spotted with groves of aspen and cotton wood, and beds of willows of ample extent.

When first seen the lake appeared smooth and polished like a vast field of glass, and took its colour from the sky which was a clear unclouded blue. It was dotted over by hundreds of pelicans white in their plumage as the fresh-fallen snow. While we yet paused, gazing rapturously upon the charmed prospect, and feasting our eyes upon its unhidden beauties, we were overtaken by a tremendous gale of wind accompanied with rain, which dissipated in a moment a lovely cottage Fancy had half constructed upon the quiet margin of the sleeping lake. Beautiful to behold is a fair young female in the soft slumber of health and innocence, but far more beautiful when startled to consciousness from her gentle rest, and bright colours chase one another across her cheeks and bosom. So with the lake, which far from losing a single attraction when roused by the wind from its repose, became even more enchanting than before; for the milk-white billows rolling like clouds over its deep blue surface seemed to add a bewitching something to the scene that did not appear to be wanting until the attention of the observer was directed to it, when it became too essential to be spared. Admonished by the storm, we dismounted from our horses, and led them in a narrow winding path, down the steep mountain side, and reaching the valley below, halted for the night at a pleasant spring near the margin of the lake. The next day we crossed a low mountain, south of the lake, to Cache Valley Creek, which we followed into a narrow defile, nearly impassable to equestrians. On either side, rose the mountains, in some places almost, and at others quite perpendicularly, to the regions of the clouds. The sun could be seen only for a short time, and that in the middle of the day. We were often compelled while struggling over the defile, to cross the stream and force our way through almost impenetrable thickets, and at times, to follow a narrow trail along the borders of precipices, where a single mis-step would inevitably have sent horse and rider to the shades of death. We saw a number of grizzly bears prowling around the rocks, and mountain sheep standing on the very verges of projecting cliffs as far above us as they could be discerned by the eye. Such was the wild and broken route which for two entire days we were obliged to pursue. We killed a grizzly bear on the evening of the eighteenth, and emerging from the mountain-pass early on the following day, came to Cache Valley, one of the most extensive and beautiful vales of the Rocky Mountain range.

This valley, called also by some, the Willow Valley, is situated about thirty miles due west of the Little Lake, from which the passage is so nearly impracticable, that it requires two days to perform the distance - at least by the route we came. It lies parallel with the Little Lake, extending nearly north and south; is sixty miles long, and fifteen to twenty broad, and is shut in on every side by lofty mountains. Numerous willow-skirted streams, that intersect and diversify it, unite and flow into Bear River, which crosses the valley, and after cutting its way through a low bald mountain, falls into the Big Lake, distant twenty miles to the west.

Cache Valley is abundantly fertile, producing every where most excellent grass, and has ever for that reason, been a favorite resort for both men and animals, especially in the winter. Indeed, many of the best hunters assert that the weather is much milder here than elsewhere, which is an additional inducement for visiting it during that inclement season. It received its name from a melancholy incident that occurred in it a few years ago. The circumstances are briefly these: -

A man in the employ of Smith, Sublette and Jackson, was engaged with a detached party, in constructing one of those subterranean vaults for the reception of furs, already described. The cache was nearly completed, when a large quantity of earth fell in upon the poor fellow, and completely buried him alive. His companions believed him to have been instantly killed, knew him to be well buried, and the cache destroyed, and therefore left him

Unknelled, uncoffined, ne'er to rise,
Till Gabriel's trumpet shakes the skies,

and accomplished their object elsewhere. It was a heartless, cruel procedure, but serves to show how lightly human life is held in these distant wilds.

In this country, the nights are cold at any season, and the climate perhaps more healthy than that of any other part of the globe. The atmosphere is delightful, and so pure and clear, that a person of good sight has been known to distinguish an Indian from a white man, at a distance of more than a mile, and herds of buffalo may be recognized by the aid of a good glass, at even fifteen to eighteen miles.

Passing down the valley, we met a number of grizzly bears, one of which of a large size, we mistook for a buffalo bull, and were only convinced of our error when the huge creature erected himself on his haunches, to survey us as we passed. These animals are of every shade of colour, from black to white, and were seen singly in the prairies, busied in digging roots, which constitute their chief subsistence until fruits ripen in the fall.

The object of our visit to Cache Valley, was to find the Free Men, but our search for them proved fruitless. We were unable to discover any recent traces either of whites or Indians, and retracing our steps, halted at the lake beneath the shade of an aged cotton wood, in the branches of which a bald eagle sat quietly on her nest, apparently indifferent to our presence, nor did she leave it during our stay. While here, we killed one of the many pelicans which were disporting on the lake, and found that it measured eight and a half feet between the tips of its extended wings.

After our return to camp, six others of the party were sent northward, on the same errand, but they were equally unsuccessful. They were absent eleven days, and saw in their route abundance of fine salt, and likewise a number of curious springs, of which a description will be given on some future page.

On the tenth day of August, a village of Shoshonees or Snake Indians, entered the valley of Bear River, fifteen or twenty miles above us, and encamped on the margin of the stream. Some of them paid us a speedy visit, and testified their friendship for us by giving us each a hearty hug. Two days after the arrival, we moved up the valley, and encamped half a mile below them. Their village consisted of about one hundred and fifty lodges, and probably contained above four hundred fighting men. The lodges were placed quite close to each other, and taken together, had much the appearance of a military camp. I strolled through it with a friend, to gratify my curiosity, as to their domestic manners. We were obliged to carry clubs, to beat off the numerous dogs, that were constantly annoying us by barking, and trying to bite our legs. Crowds of dirty naked children followed us from lodge to lodge, at each of which were seen more or less filthy but industrious women, employed in dressing skins, cutting meat into thin strips for drying, gathering fuel, cooking, or otherwise engaged in domestic labour. At every lodge, was a rack or frame, constructed of poles tied together, forming a platform, covered over with half-dried meat, which was curing over a slow fire. The women were all at work, but not so the men. Half of them were asleep in the lodges, and the rest either gaming, keeping guard over their horses, or leisurely strutting about camp. They are extremely jealous of their women, though I could not help thinking, with but slight occasion, when I surveyed the wrinkled, smoke-dried unprepossessing features of the latter, and the dirt and filth by which they are surrounded. Cupid must have a queer taste, if he can find marks for his arrows among the she snakes of this serpent tribe. We spoke to several of them, but they either feigned not to hear, or retired at once. After gratifying our curiosity, which did not require long, we purchased a few buffalo robes, and skins of other kinds, for trifles of little value to us, yet by them prized highly, and returned sadder though wiser to our own encampment.

Most of the Rocky Mountain Indians are given to prigging, as we have already had a taste of proof. The Snakes are by no means deficient in this accomplishment, and at almost every visit they made to us many little articles acquired a trick of vanishing with the most marvelous dexterity. However we left them on the sixteenth, and returned to Ham's Fork, by way of a small stream, called Muddy, from the turbid appearance of its waters. This little stream rises against Ham's Fork, and flows south of west thirty miles, emptying into Bear River, nearly opposite to the spring which marks the pass to the head of Little Lake. It is noted as being the best route from Ham's Fork to Bear River, there being no steep ascents or descents in the whole distance.

On Ham's Fork we cached our goods, and separated into three parties, headed respectively by Messrs. Fontenelle, Dripps, and Robidoux, who had each his portion of hunting ground specified, in order to avoid interference with the rest. Mr. Fontenelle was to hunt to the southward on the western tributaries of Green River; Mr. Dripps to the northeast on the sources of the same stream, and Mr. Robidoux northward on the head waters of Lewis River.

We separated on the twenty-third, and departed in quest of adventures and beaver, - my unlucky stars having induced me to join Mr. Fontenelle's party, which met with the least of either. We rambled about in the Eut mountains, on the sources of Black's Fork, and Henrie's Fork, explored them to their outlets, and returned to the caches after a month's absence, having starved one half of the time. After leaving Ham's Fork, we saw no buffalo until our return, and killed no game of any kind, except one elk, two or three goats, and a few beaver. We were nicely frightened by a party of Crow Indians, who crawled up to our encampment one dark night, and fired a volley over our heads. We sprang to our feet, but before we could return the compliment, they came into camp shouting Ap-sah-ro-ke, - Ap-sah-ro-ke, (Crows,) and laughing heartily at the confusion their novel manner of introducing themselves had occasioned us. From them we ascertained that the Free Men who had caused us so much unavailing search, were on the Yellow Stone River. Two of our men were sent with the Crows, to raise the cache on Sweet Water, proceed with them to their village, and trade until further orders. Previous to their departure, the Crows gave us a few practical lessons in the art of pilfering, of which they are the most adroit and skilful professors in all this region, if not the world. No legislative body on earth ever made an appropriation with half the tact, facility, and success, that characterize these untaught sons of the forest.

Chapter X

On the twentieth of September, five of us left the party to 'hunt' several small streams in the vicinity of Bear River. We proceeded to the mouth of the Muddy, and followed Bear River down fifteen miles to the mouth of Smith's Fork, where we saw recent traces of brother trappers and Indians. The same evening I was thrown from my horse, by which my gun was broken so as to render it entirely useless. The feelings of a trapper may better be imagined than described, after losing his only means of subsistence and defence, in hourly danger of his life and thrown entirely upon the charity of his comrades, from whom should he get accidentally separated, he must either perish miserably, or suffer privations and agonies compared to which death were mercy, before he could find the company. From Smith's Fork we passed down to Talma's Fork, - so named in honour of the great French tragedian, - eight miles below. The plains of this stream as also those of Bear River, were covered with buffalo, one of which we killed, and after packing the meat travelled up the fork fifteen miles into the mountains, where it divides into three branches of nearly equal size. On the middle one of these branches two miles above the fork, we found a large quantity of beautiful white salt, formed by the total evaporation of a pond, on the rocks forming the bed of which it was encrusted. From this point we passed up to the head of the western fork, and thence crossed to a small stream called Beaver Creek, from the uncommon labours of those industrious animals, which are here observed, forming a succession of dams for several miles. We first tasted the waters of the Columbia river which has its source in this little stream, on the first of October, after which we continued our hunt down the creek to its mouth, twenty miles from its fountain head, and all the way confined between high mountains. The narrow bottoms along it were occasionally covered with bushes bearing a delicious fruit called service berries, by the American hunters, and pears (Des Poires) by the Canadians: a species of black hawthorn berries, wild currents, goose berries, black cherries, and buffalo berries were also at intervals abundant. At the mouth of Beaver Creek the mountains retire apart leaving a beautiful valley fifteen miles long, and six to eight broad, watered by several small streams which unite and form "Salt River," so called from the quantities of salt, in a chrystalized form, found upon most of its branches. At the northern or lower end of the valley we observed a white chalk-like appearance, which one of our party, (who had been here with others in quest of the Free men) recognized to be certain singular springs. His account of them excited my curiosity and that of one of my companions, so much that we determined upon paying them a visit. With this intention we set out early one fine morning and reached our place of destination about noon, after an agreeable ride of three or four hours.

We found the springs situated in the middle of a small shallow stream, in the open level prairie. Rising from the middle of the brook, were seen seven or eight semi globular mounds self-formed by continual deposites of a calcarious nature, which time had hardened to the consistency of rock. Some of them were thirty or forty feet in circumference at the base, and seven or eight feet high. Each of them had one or more small apertures (similar in appearance to the mouth of a jug) out of which the water boils continually, and these generally, though not invariably, at the top of the mound. The water that boils over, deposits continually a greenish, slimy, foeted cement, externally about the orifices, by constant accretions of which, the mounds are formed. The water in these springs was so hot, that we could not bear our fingers in it a moment, and a dense suffocating sulphurous vapour is constantly rising from them. In the bases of the mounds, there were also occasional cavities from which vapour or boiling water was continually emitted. Some of the mounds have long since exploded, and been left dry by the water. They were hollow, and filled with shelving cavities not unlike honey-comb. These singular springs are known to the Rocky Mountain hunters by the name of the Boiling Kettles, and are justly regarded as great curiosities. After spending a couple of hours very agreeably in examining these remarkable fountains, we returned to camp, well satisfied for the fatigue of thirty miles' travel, by the opportunity we had enjoyed of perusing one of the most interesting pages of the great book of nature. A fair day and a beautiful prospect, enhanced the pleasure and reward of our excursion.

Leaving the valley, we returned slowly back to Talma's Fork, trapping many small streams by the way, near some of which we saw considerable deposites of pure salt. We had a severe storm of rain, on the twenty-third, which finally changed to snow. Except occasional light showers, this was the only interruption to fair weather that we had experienced since we left the caches. On the twenty-seventh, we were greatly alarmed by one of the party (Milman) returning at full speed from a visit to his traps, and yelling in tones of trepidation and terror, that fear had rendered less human than the screams of the panther. We sprung to our arms, rushed our horses into camp, and awaited his approach with feelings wrought up to the highest pitch of excitement between suspense and apprehension. As he approached nearer, however, his voice becoming less unearthly, at length relaxed into something like human speech; and guessing at his meaning, rather by the probability of the case, than by any actual sounds he uttered, we made out the words "Indians! Indians!" The lapse of a few moments brought him up, exclaiming, "Boys, I am wounded!" We saw at once that a well-directed ball had been intercepted by his gun, which thus evidently saved his life. The ball had been cut into several pieces by the sharp angle of the barrel, one of which, glancing off, had lodged in the fleshy part of his thigh. The same bullet, previous to striking his gun, had passed through the neck of his mule, and grazed the pommel of his saddle. He was also struck in the shoulder, by an arrow, but both wounds were slight.

After recovering his wonted control over the faculties of speech, he gave us the following particulars of the affair, which was ever afterwards facetiously termed "Milman's Defeat." Whilst jogging along, three or four miles from camp, and calculating the probable sum total of dollars he should accumulate from the sales of furs he purposed taking from his traps that morning, his dog suddenly commenced barking at some invisible object which he supposed to be a squirrel, badger, or some other small animal, that had taken refuge in its burrow. Satisfied of his own sagacity in arriving at this conclusion, he advanced thoughtlessly, until he reached the top of a gently - ascending knoll, whence, to his utter astonishment and dismay, he discovered the heads of seven or eight Indians, peeping ferociously up from a patch of sage, not thirty steps beyond him, and at the same instant three guns were fired at him, by way of introduction. This sort of welcome by no means according with his notions of politeness, he wheeled about with the intention of making his stay in the vicinity of persons whose conduct was so decidedly suspicious, as brief as possible. His mule seemed however far less disposed to slight the proffered acquaintance, and positively refused to stir a single peg. In the meantime, the Indians starting up, showered their compliments in the shape of arrows upon him with such hearty good will, that he was forced to dismount, intending to return their kindness with an impromptu ball from his rifle; but ere he could effect this, the Indians, divining his purpose, and overcome by so touching a proof of friendship, bowed, scraped, and retired precipitately, in all likelihood to conceal their modest blushes at his condescension. Just then, too, madam Long Ears, probably resenting their unceremonious departure, betrayed symptoms of such decided displeasure, that Milman was induced perforce, to remount, after he had withdrawn an arrow from his shoulder, but before he had accomplished his purpose of presenting the red-skins with the contents of his gun, free gratis, in exchange for their salute; and he was borne away from the field of his achievements with a gallantry of speed that would not have discredited the flight of Santa Anna from the battle-plain of San Jacinto, but which Long Ears had never displayed, unless fear lent the wish of wings to her activity. Milman did not, he said, discover that he had been struck by a ball, until he saw the blood, which was just before he reached camp.

Shortly after the return of Milman, two Indians, to our surprise, came coolly marching up to camp, who proved, on their approach, to be Snakes, a young savage and his squaw. They had left their village at the mouth of Smith's Fork, for the purpose of hunting big-horns, (Rocky Mountain sheep,) in a mountain near by, from which he discovered us. We questioned him until we were perfectly satisfied that he was an innocent, harmless fellow, and in no way associated with the party which had fired upon Milman, though we strongly suspected them to be Snakes. He soon took his leave, and shortly disappeared in the forest of pines, which encircle the bases of all lofty mountains in this region. We departed also, not doubting but that the Indians who attacked Milman, would hang about, seeking other opportunities to do us injury.

Passing up the east fork of the three, into which Talma's Fork is subdivided, we crossed it and ascended a high mountain eastward, on the summit of which we halted at midnight, and, having tied our beasts to cedars, of which there were a few scattered here and there, threw ourselves down to sleep, almost exhausted with fatigue, and still haunted by fears of murdering savages, who might have dogged our footsteps, and be even now only waiting the approach of dawn to startle us with their fiendish yells and arrows, and take our - scalps. At day break we resumed our weary march, forced our way, though with great difficulty through a chaos of snow banks, rocks and fallen pines, to the east side of the mountain, and at last descended to the source of Ham's Fork, on which we passed the night. The next day we reached an open valley of considerable extent, decked with groves of aspen, and beds of willows, and grazed by a numerous herd of buffalo. Midway of this valley, on the western side, is a high point of rock, projecting into the prairie and overlooking the country to a great distance. Imagine our surprise when we beheld a solitary human being seated on the very pinnacle of this rock, and apparently unconscious of our approach, though we were advancing directly in front of him, - and he so elevated that every object however trifling, within the limit of human vision seemed to court his notice; and what made it still more singular, there was evidently no person in or near the valley except ourselves. We halted before him, at a short distance, astonished to see one solitary hero, who seemed to hide himself from he knew not what - friends or foes; but firm as the giant rock on which he sat as on a throne, seemed calmly to await our approach, then to hurl the thunder of his vengeance upon us, or fall gloriously like another Warwick, disdaining to ask what he can no longer defend. With mingled feelings of respect and awe we approached this lord of the valley, gazed admiringly up at the fixed stolidity of his countenance, and lo! he was dead.

I afterwards learned that this Indian was taken in the act of adultery with the wife of another, and put to death by the injured husband. He was a Shoshone, and was placed in this conspicuous position by the chief of the tribe, as a warning to all similar offenders.

On the thirty first we reached the caches where we found Robidoux with a small party of men. Fontenelle and Dripps, together with the Free Men, and a detachment of a new company, styled the Rocky Mountain Fur Company, were all in Cache Valley, where they intended to establish their winter quarters. Robidoux remained here twelve days, awaiting promised assistance from Fontenelle, to aid him in transporting the goods to Cache Valley. At the end of that time, impatient of their slow coming, and admonished by the more rapid approach of starvation which was already grinning at us most horribly, he resolved to re-cache a part of the goods, and start with the balance.

We set off in the midst of a severe snow-storm, accompanied with chilling winds, which blew directly in our faces, and, having braved with the best temper we could, a whole day of such exposure, encamped at evening on the margin of Muddy Creek. We were met next day at noon, by the expected party. They continued on to raise the cache we had left, whilst we journeyed down to the mouth of the Muddy, there to await their return. In the meantime, hunters were dispatched in pursuit of game, who brought back with them, at the expiration of two days, the flesh of several fine bulls.

The report of Milman's defeat, was received in Cache Valley, from a party of Snakes some time before we arrived, with the additional information, that the young Indian who paid us a visit on that memorable morning, was killed on the evening of the same day, and his wife taken prisoner, though she escaped the night following.

On the third day after we reached Bear River, the party dispatched for that purpose, returned with the contents of the cache, and on the fifth we arrived in sight of the camp, exchanged salutes, and hastened to grasp the honest hands of our hardy old comrades, glad to meet and mingle with them again after a long absence, and listen to their adventures, or recount our own.

Chapter XI

We remained about ten days in the northern point of Cache Valley, in a small cove frequently called Ogden's Hole, in compliment to a gentleman of that name of the Hudson Bay Company, who paid it a visit some years since. Meanwhile, the men amused themselves in various ways, - drinking, horse racing, gambling, etc. and at the same time, Mr. J. H. Stevens, an intelligent and highly esteemed young man, gave me the following account of his adventures with Robidoux, which was confirmed by others of the party.

"After leaving you," said he, "we trapped Ham's Fork to its source, crossed over to Smith's Fork, and there fell in with a party of Iroquois, who informed us that Smith, Sublette and Jackson, three partners who had been engaged in the business of this country for some years past, had sold out to a new firm, styled the Rocky Mountain Fur Company. This arrangement was made on Wind River, a source of the Big Horn, in July of last year. From that place parties were sent out in various directions, amongst which was one led by Fraeb and Jarvis, consisting of twenty two hired men, and ten free Iroquois, with their wives and children - which departed to hunt on the waters of the Columbia. The Iroquois, however, became dissatisfied with some of the measures adopted by the leaders of the party, and separated from them to hunt the tributaries of Bear River, where we found them. Robidoux engaged three of them, and the others promised to meet us in Cache Valley, after the hunting season. One of those hired, was immediately despatched in pursuit of Dripps, who joined us at the Boiling Kettles, on Salt River, from whence we proceeded to its mouth, and there fell in with Fraeb and Jarvis. Arrangements were now made for both companies to hunt together, and we travelled thence sixty miles to the mouth of Lewis River, and down Snake river eighty or ninety miles to Porteneuf. Here we cached our furs, and thence continued down Snake River to the falls, forty or fifty miles below the mouth of Porteneuf. These falls are a succession of cascades by which the river falls forty or fifty feet in a few rods. "At the Falls we separated into two parties, one of which was to hunt the Cassia, and other streams in the vicinity, whilst the other, consisting of twenty two men, myself included, was sent to the Maladi. Our party left Snake River, and travelled north of west, through a barren desert, destitute of every species of vegetation, except a few scattering cedars, and speckled with huge round masses of black basaltic rock. At noon, we entered on a tract of country entirely covered with a stratum of black rock, which had evidently been in a fluid state, and had spread over the earth's surface to the extent of forty or fifty miles. It was doubtless lava, which had been vomited forth from some volcano, the fires of which are now extinct.

"We proceeded on over this substance, hoping to cross the whole extent without difficulty, but soon met with innumerable chasms, where it had cracked and yawned asunder at the time of cooling, to the depth often of fifty feet, over which we were compelled to leap our horses. In many places the rock had cooled into little wave-like irregularities, and was also covered with large blisters, like inverted kettles, which were easily detached by a slight blow. One of these was used as a frying pan, for some time afterwards, and found to answer the purpose quite well. In the outset of our march over this bed of lava, we got along without much trouble, but were finally brought to a full stop by a large chasm too wide to leap, and forced to return back to the plain. At this time we began to feel an almost insupportable thirst. The day was an excessively sultry one, and the lava heated to that degree that we were almost suffocated by the burning atmosphere, that steamed up from it. We had, moreover, lived for some time past, upon dried buffalo meat, which is alone sufficient to engender the most maddening desire for water, when deprived of that article.

"One or two individuals, anticipating the total absence of any stream or spring on the route, had providently supplied themselves with beaver skins of water, previous to our departure in the morning, but this small supply was soon totally exhausted. At dark we found ourselves involved in a labyrinth of rocks, from which we sought, without success to extricate ourselves, and were finally obliged to halt and await the rising of the moon. Meantime we joyfully hailed the appearance of a shower, but greatly to our chagrin, it merely sprinkled slightly, and passed over. However it was not entirely lost, for we spread out our blankets and eagerly imbibed the dampness that accumulated, but the few drops thus obtained, provoked rather than satisfied the wild thirst that was raging within us.

"At the expiration of a couple of hours, the moon rose, and we proceeded cautiously in the direction of a blue mountain, where we conjectured that the river Maladi took its rise. Through the rest of the night we toiled on, and at length we saw the sun climbing the east. But the benefit of his light was a mere feather in the scale, compared with the double anguish occasioned by the added heat. Some of the party had recourse to the last expedient to mitigate their excessive thirst, and others ate powder, chewed bullets, etc. but all to no purpose. At eight o'clock, we reached a narrow neck of the rock or lava, which we succeeded in crossing. Some of our companions explored the interior of frightful chasms in search of water, but returned unsuccessful. Subordination now entirely ceased. Every one rushed forward without respect to our leaders, towards a rising plain which separated us from the blue mountain which had been our guiding beacon since the night. On reaching the summit of the plain, the whole valley about the mountain presented a sea of rock, intersected by impassable chasms and caverns.

"Orders were now given for every one to shift for himself, and exercise his best judgement in the endeavour to save his life. One of the men immediately turned his horse from north west, which had been thus far our course, to the north east, and declared that if any thought proper to follow him, they would be rewarded by the taste of water before night. We all followed him, rather because the route seemed less difficult, than from any well-grounded hope of realizing his promise.

"Our suffering became more and more intense, and our poor animals, oppressed with heat and toil, and parching with thirst, now began to give out, and were left by the way side. Several of our poor fellows were thus deprived of their horses, and though almost speechless and scarcely able to stand, were compelled to totter along on foot. Many of our packed mules, unable to proceed any further, sank down and were left with their parched tongues protruding from their mouths. Some of the men too, dropped down totally exhausted, and were left, beseeching their companions to hasten on, and return to them with water, if they should be so fortunate as to succeed in reaching it.

"At length, when all were nearly despairing, and almost overcome, one of our companions who had outstripped us to the top of a hill, fired off his gun. The effect was electrical. All knew that he had found water, and even our poor beasts understood the signal, for they pricked up their drooping ears, snuffed the air, and moved off at a more rapid pace. Two or three minutes of intense anxiety elapsed, we reached the top of the hill, and then beheld what gave us infinitely more delight than would the discovery of the north west passage, or the richest mine of gold that ever excited, the cupidity of man.

"There lay at the distance of about four miles, the loveliest prospect imagination could present to the dazzled senses - a lovely river sweeping along through graceful curves. The beauteous sight lent vigour to our withered limbs, and we pressed on, oh! how eagerly. At sunset we reached the margin of the stream, and man and beast, regardless of depth, plunged, and drank, and laved, and drank again. What was nectar to such a draught! The pure cool reviving stream, a new river of life, - we drank, laughed, wept, embraced, shouted, - and drank, shouted, embraced, wept, and laughed again. Fits of vomiting were brought on by the excessive quantities we swallowed, but they soon passed off, and an hour or so saw us restored to our usual spirits.

"We spent that night and the following morning in the charitable office of conveying water to our enfeebled companions, who lingered behind, and the poor beasts that had been also left by the way, and succeeded in getting them all to camp, except the person and animals of Charbineau,* one of our men, who could nowhere be found, and was supposed to have wandered from the trail and perished.

*This was the infant who, together with his mother, was saved from a sudden flood near the Falls of the Missouri, by Capt. Lewis, - vide Lewis and Clark's Expedition. [W. A. F.]

Chapter XII

"Next morning," continued Stevens, "several successive reports of firearms were heard apparently at the distance and direction of a mile or so below camp. Supposing the shots to have been fired by Charbineaux, one of our men was despatched in quest of him, but he shortly after returned, accompanied by several trappers who belonged to a party of forty, led by a Mr. Work, a clerk of the Hudson Bay Company. These men were mostly half breeds, having squaws and children. They live by hunting furred animals, the skins of which they dress and exchange for necessaries at the trading posts of that company, on the Columbia and its tributaries.

"Two days before we met them, five of their hunters were fired upon by a party of Indians, who lay concealed in a thicket of willows near the trail. One of them was killed on the spot, and a second disabled by a shot in the knee. An Indian at the same moment sprang from the thicket and caught the wounded man in his arms, who, well knowing that torture would be the consequence of captivity, besought his flying comrades to pause and shoot either the Indian or himself. Heeding his piteous cry, one of the retreating hunters, more bold, or more humane than the other two, wheeled and fired, but missed his aim, and hastily resumed his flight. The exasperated savage, at this, let go his hold, pursued, overtook, and killed the unlucky marksman, while the wounded man crept into a thicket and effectually concealed himself till night, when he made his escape. The bodies of the two dead men were found the next day; both had been stripped and scalped. Beside one of them lay a gun, broken off at the breech, and charged with two balls without powder. They were buried as decently as circumstances would permit, and the place of interment carefully concealed to prevent their last repose being rudely disturbed by the Indians, who frequently, with a fiendish malice, tear open the graves of their victims, and leave their bones to bleach upon the soil.

"The river on which we were now encamped, and the fortunate and timely discovery of which had saved us from the last extremity of thirst, is called 'La Riviere Maladi,' (Sick River,) and owes its name to the fact that the beaver found upon it, if eaten by the unwary hunter, causes him to have a singular fit, the symptoms of which are, stiffness of the neck, pains in the bones, and nervous contortions of the face. A party of half-starved trappers found their way to this stream a few years since, and observing plenty of beaver 'signs,' immediately set their traps, in order to procure provisions. At dawn the next day, several fine large fat beavers were taken, and skinned, dressed and cooked, with the least possible delay. The hungry trappers fed ravenously upon the smoking viands, and soon left scarce a single bone unpicked. Two or three hours elapsed, when several of the party were seized with a violent cramp in the muscles of the neck; severe shooting pains darted through the frame, and the features became hideously convulsed. Their companions were greatly alarmed at their condition, and imagined them to be in imminent danger. However, at the expiration of an hour, they were quite recovered, but others had meantime been attacked in the same way. These also recovered, and by the following morning all had passed the ordeal, save one, who having escaped so much longer than the rest, fancied himself entirely out of danger, and indiscreetly boasted of his better constitution, laughing at what he called the effeminacy of his companions.

"During the very height of his merriment, which by the way, was any thing but agreeable to his comrades, he was observed to turn pale, his head turned slowly towards his left shoulder, and became fixed, his mouth was stretched round almost to his ear on the same side, and twitched violently, as if in the vain endeavor to extricate itself from so unnatural a position, and his body was drawn into the most pitiable and yet ludicrous deformity. His appearance, in short, presented such an admirable and striking portraiture of the 'beautiful boy,' that his companions could not help indulging in hearty peals of laughter at his expense, and retorted his taunts with the most provoking and malicious coolness. When he recovered he was heard to mutter something about 'whipping,' but probably thought better of it afterwards, as he never attempted to put his threat into execution. Indeed, he subsequently acknowledged that he had been justly treated, and was never, from that time forth, heard to speak of his 'constitution.'

"Notwithstanding that we were well aware of these facts, we could not resist the temptation of a fine fat beaver, which we cooked and eat. But we were all sick in consequence, so much so, in short, that I do not believe a single one of us will ever be induced to try the same experiment again, no matter how urgently pressed by starvation."

There is a small stream flowing into the Big Lake, the beaver taken from which, produce the same effect. It is the universal belief among hunters, that the beaver in these two streams feed upon some root or plant peculiar to the locality, which gives their flesh the strange quality of causing such indisposition. This is the only mode in which I ever heard the phenomena attempted to be explained, and it is most probably correct.

"We trapped the Maladi to its source, then crossed to the head of Gordiaz River, and trapped it down to the plains of Snake River, from whence we returned to Cache Valley by the way of Porteneuf, where we found Dripps and Fontenelle, together with our lost companion Charbineaux. He states that he lost our trail, but reached the river Maladi after dark, where he discovered a village of Indians. Fearing that they were unfriendly, he resolved to retrace his steps, and find the main company. In pursuance of this plan, he filled a beaver skin with water, and set off on his lonely way. After eleven day's wandering, during which he suffered a good deal from hunger, he attained his object, and reached the company at Porteneuf. The village he saw was the lodges of the Hudson Bay Company, and had he passed a short distance below, he would have found our camp. But his unlucky star was in the ascendant, and it cost him eleven day's toil, danger, and privation to find friends."

Such was the narrative Mr. Stevens gave me of the adventures of Robideaux's party.

Chapter XIII

From Ogden's Hole, we passed by short marches down Cache Valley forty miles to Bear river, where we remained at the same encampment a whole month. During this time it stormed more or less every day, and the snow accumulated to such a depth that four of our hunters, were compelled to remain away from camp for thirty four days, the impossibility of travelling having prevented their return from an expedition after game. In all December the snow lay upwards of three feet deep, throughout Cache Valley; in other parts of the country the depth was still greater. In the latter part of this month, we separated from Fraeb and Jarvis, and crossed over to the Big Lake, a distance of thirty miles which we accomplished in four days. The "Big Lake" is so called in contra-distinction to the Little Lake, which lies due East from it fifty miles, and which has been described in a former chapter. It is sometimes also called "Salt Lake," from the saline quality of its waters. An attempt has been recently made to change the name of this lake to Lake Bonnyville, from no other reason that I can learn, but to gratify the silly conceit of a Captain Bonnyville, whose adventures in this region at the head of a party, form the ground work of "Irving's Rocky Mountains." There is no more justice or propriety in calling the lake after that gentleman, than after any other one of the many persons who in the course of their fur hunting expeditions have passed in its vicinity. He neither discovered, or explored it, nor has he done any thing else to entitle him to the honour of giving it his name, and the foolish vanity that has been his only inducement for seeking to change the appellation by which it has been known for fifty years, to his own patronymic, can reflect no credit upon him, or the talented author who has lent himself to the service of an ambition so childish and contemptible.

The dimensions of the Big Lake have not been accurately determined, but it may be safely set down as not less than one hundred miles in length, by seventy or eighty broad. It was circumnavigated a few years since by four men in a small boat, who were absent on the expedition forty days, and on their return reported that for several days they found no fresh water on its western shore, and nearly perished from the want of that necessary article. They ascertained that it had no visible outlet, and stated as their opinion that it was two hundred miles long and one hundred broad, but this was doubtless a gross exaggeration. I ascended a high mountain between Bear River and Webber's Fork, in order to obtain an extensive view of it, but found it so intersected by lofty promontories and mountains, not only jutting into it from every side, but often rising out of its midst, that only thirty or forty square miles of its surface could be seen. Its waters are so strongly impregnated with salt that many doubt if it would hold more in solution; I do not however think it by any means saturated, though it has certainly a very briny taste, and seems much more buoyant than the ocean. In the vicinity of the Big Lake we saw dwarf oak and maple trees, as well on the neighboring hills as on the border of streams. This was the first time since leaving the Council Bluffs that we have seen timber of that description.

About the first of February we ascertained that a number of Caches we had made previous to our leaving Cache Valley, had been robbed by a party of Snakes, who without doubt discovered us in the act of making them. However the "Horn Chief," a distinguished chief and warriour of the Shoshonee tribe, made them return every thing he could find among them into the Caches again, though a multitude of small articles to the value of about two hundred dollars were irrecoverably lost. I had almost forgotten to record a debt of gratitude to this high souled and amiable chief for an act of chivalry that has scarce a parallel in the annals of any age or nation, in respect either of lofty courage, or disinterested friendship. The Horn Chief is noted for his attachment to the whites, numbers of whom owe to him not only the protection of their property, but the safety even of their lives. He is the principle chief of the Snakes, and forms a striking contrast to his people, being as remarkable for his uprightness and candour as they are noted for treachery and dishonesty.

While we remained near the Snake village on Bear River, the preceding autumn, they formed a plot to massacre us solely for the purpose of possessing themselves of our arms and baggage. Relying on their professions of friendship, and unsuspicious of ill faith, we took no precautions against surprise, but allowed them to rove freely through camp, and handle our arms, and in short gave them every advantage that could be desired. The temptation was too much for their easy virtue. Such an opportunity of enriching themselves, though at the cost of the blackest ingratitude, they could not consent to let slip, and therefore held a council on the subject at which it was resolved to enter our camp under the mask of friendship, seize our arms, and butcher us all on the spot.

In these preliminary proceedings the Horn Chief took no part, he having preserved the strictest silence throughout the whole debate. But when the foul scheme was fully resolved upon and every arrangement made for carrying it into effect, he arose and made a short speech in which he charged them with ingratitude, cowardice, and the basest breach of faith, and after heaping upon them the most stinging sarcasms and reproaches, concluded by telling them he did not think they were manly enough to attempt putting their infamous design into execution, but to remember if they did, that he would be there to aid and die with those they purposed to destroy.

Early the following morning the Snakes assembled at our camp with their weapons concealed beneath their robes; but this excited no suspicion for we had been accustomed to see them go armed at all times and upon every occasion. None of their women or children however appeared, and this was so unusual that some of my companions remarked it at the time; still the wily devils masked their intention so completely by an appearance of frank familiarity and trusting confidence, that the idea even of an attack never occurred to us.

At length when they had collected to more than thrice our number, the Horn Chief suddenly appeared in the centre of our camp, mounted on a noble horse and fully equipped for war. He was of middle stature, of severe and dignified mien, and wore a visage deeply marked by the wrinkles of age and thought, which with his long gray hairs showed him to have been the sport of precarious fortune for at least the venerable term of sixty winters. His head was surmounted by a curious cap or crown, made of the stuffed skin of an antelope's head, with the ears and horns still attached, which gave him a bold, commanding, and somewhat ferocious appearance.

Immediately upon his arrival he commenced a loud and threatening harrangue to his people, the tenor of which we could not comprehend, but which we inferred from his looks, tone of voice and gestures, boded them no good, and this opinion was strengthened by their sneaking off one after another until he was left quite alone. He followed immediately after, himself, leaving us to conjecture his meaning. However he afterwards met with the Iroquois, and informed them of the whole matter, and the same time showing the tip of his little finger, significantly remarked that we escaped "that big."

It appears they were assembling to execute their diabolical plot, and about to commence the work of blood when the Horn Chief so opportunely arrived. He instantly addressed them, reminded them of his resolution, dared them to fire a gun, called them cowards, women, and in short so bullied and shamed them that they sneaked away without attempting to do us any injury. It was not for months afterwards that all this came to our knowledge and we learned how providential had been our deliverance, and how greatly we were under obligation to the friendship, courage, and presence of mind of this noble son of the forest, whose lofty heroism in our defence may proudly rival the best achievements of the days of chivalry.

Some days after the robbery of the Caches, seventeen horses were stolen from a detachment of our party which had been sent to Cache Valley for provisions. The were about sixty in number, and supposed to be Blackfeet. They departed in the direction of Porteneuf. This misfortune prevented our obtaining supplies of meat, and we were consequently reduced to the necessity of living on whatever came to hand. Famished wolves, ravens, magpies, and even raw hide made tender by two days boiling, were greedily devoured. We lived or rather starved in this manner ten or twelve days, daily expecting the arrival of our hunters with meat, but they came not, and we were compelled to return to Cache Valley where we halted on the first of March on Cache Valley Creek. We saw in our route several boiling springs, the most remarkable of which bursts out from beneath a huge fragment of rock, and forms a reservoir of several rods in circumference, the bottom of which was covered with a reddish slimy matter. The waters of these springs was as hot as in those on Salt River. They are situated near the trail that leads from the head of Cache Valley to the Big Lake.

We found the snow in Cache Valley reduced to the depth of eighteen inches, but covered with a crust so thick and firm that it cuts our horse legs, making them bleed profusely, and the trail of our poor beasts was sprinkled with blood at every step, wherever we went.

During the month of March, we proceeded slowly to Bear River, starving at least one half the time. Our horses were in the most miserable condition, and we reduced to mere skeletons. Our gums became so sore from eating tough bull meat, that we were forced to swallow it without chewing; and to complete our misery, many of us were nearly deprived of sight from inflammation of the eyes, brought on by the reflection of the sunbeams on the snow.

Early in April wild geese began to make their appearance, - a happy omen to the mountain hunter. The ice soon disappeared from the river, and the days became generally warm and pleasant, though the nights were still extremely cold. About this time three Flathead Indians came to us from the Hudson Bay trappers, who had passed the winter at the mouth of Porteneuf, and reported that the plains of Snake River were already free from snow. This information decided our leaders to go there and recruit our horses preparatory to the spring hunt, which would commence as soon as the small streams were disencumbered of their icy fetters; and we set about the necessary arrangements for departure.

Chapter XIV

On the fourth of April, having cached our furs, and made other necessary arrangements for our journey, we set off, and proceeding but slowly, though with great fatigue, owing to the great depth and hardness of the snow, which though encrusted stiffly, would by no means bear the weight of our horses, - accompanied but a few miles, when we halted for the night at a spring source, in the northern extremity of Cache Valley. The following day we crossed a prairie hill, and encamped at evening at the fountain source of the south fork of Porteneuf, having seen on our route great numbers of buffalo, and many with young calves. We found the snow next day increased to the depth of from three to five feet, and floundered along through it for a few miles, though with the greatest toil and difficulty. Buffalo were quite as numerous on this day as the preceding, and we caught thirty or forty of their calves alive in the snow. Quite as many more were observed either killed or maimed by the frighted herds in their fugitive course. We rested that night on the south side of a hill, which the wind had partly denuded of snow, leaving here and there spots quite divested of it; but found neither grass nor water, both of which were greatly needed, and but scant supply of sage (wormwood) which we were obliged from the absence of every other, to use as a substitute for fuel. Water we obtained both for ourselves and horses, by melting snow in our kettles, - a tedious and vexatious process. The exertion of another day sufficed us to reach a point where fuel was more abundant, and of somewhat better quality, - a few scattered clusters of large willows furnishing us the necessary firewood to our comfort. We killed a number of buffalo, but upon any one of them the least particle of fat could not be found, and our fare was therefore none of the best, as may well be imagined. On the morning of the eighth, several men were sent out with directions to drive, if possible, a herd of bulls down the river. Could this have been effected, we should have had a tolerable road for our feeble horses to follow, but no such good fortune was the reward of our endeavours, for the buffalo refused absolutely to move, and were all, to the number of fifty and upwards, killed on the spot. Disappointed of this hope, we had no alternative but to resume our weary march as before, which we did at sunrise, on the following day. At first, we got along with tolerable ease, but as the forenoon advanced, the warmth of the sun so melted and softened the crust of the snow, that our horses plunged in at every step, and speedily became quite exhausted from the excessive fatigue of constantly breaking through, and forcing their way under such disadvantage. There was no alternative but for us to carry them, since they could neither carry us nor themselves even, and we therefore procured poles, and transported them two miles through the snow to a hill side, which was accomplished only at the cost of incredible labour, hardship and misery. In addition to this, we had our baggage, which lay scattered along the whole distance, from one encampment to the other, to collect and bring in on our shoulders, a work of immense toil, as at almost every step the crust gave way and engulphed us up to our armpits in the damp snow. However, we had the pleasure of seeing everything safe at camp in the evening, except three or four of our poorest horses, which being unable to extricate from the snow, we were obliged to abandon to their fate.

All hands were employed on the tenth, in making a road. We marched on foot one after another in Indian file, ploughing our way through the snow to the forks of Porteneuf, a distance of six miles, and back again, thus beating a path for our horses, the labour of which almost overcame our strength. When we returned, many of us were near dropping down from fatigue, so violent had been our exertion. At three o'clock next morning, rousing our weary limbs and eyelids from their needful rest and slumber, we pursued our journey, and succeeded in reaching a narrow prairie at the forks, which was found nearly free from snow. Here we remained over the following day, to refresh our half-dead horses, and rest our nearly exhausted selves. In the mean time we were visited by two hunters of the Hudson Bay Company, who gave us the grateful information that our troubles were nearly at an end, as the snow entirely disappeared after a few miles below. Six miles marching on the thirteenth, brought us out of the narrow defile we had hitherto with so much labour threaded, and into a broad and almost boundless prairie, which far as the eye could reach, was bare, dry, and even dusty. The sensation produced by this sudden transition from one vast and deep expanse of snow which had continually surrounded us for more than five months, to an open and unincumbered valley of one hundred miles in diameter, over which the sun shed its unclouded warmth, and where the greenness of starting verdure gladdened the eye was one of most exquisite and almost rapturous pleasure. Our toils were past, our hardships were over, our labours were at an end, and even our animals seemed inspired with fresh life and vigour, for they moved off at a gallop, of their own accord, evidently delighted to find their feet once more on terra firma. Since our departure from Cache Valley to this point, where Porteneuf leaves the mountains, we have made a distance of sixty miles, which to accomplish, has cost us nine day's toil.

We moved leisurely down Porteneuf on the following day, a distance of four miles, and came upon the camp of Mr. Work with the Hudson Bay trappers, (who it will be remembered, were met by Robidoux's party in the fall), with whom we pitched our quarters. From these people we procured some excellent dried meat, which, having been cured and prepared in the fall, when the buffalo were in good condition, was really most welcome; and of which we partook heartily, believing, half-famished as we were, that more delicious food never feasted the appetite of man. In consequence of a storm of sleet that lasted for two days we were forced to remain here until the morning of the seventeenth, when the sun re-appeared, and we departed, though owing to a quarrel Mr. Works had with one of his men, which resulted in the fellow joining our party, not until we had narrowly escaped flogging the "Nor'westers" as the Hudson Bay people are sometimes called, since the junction with that company of one called the Northwest Fur company, of recent date. One of Mr. Work's people, foolishly imagining that Mr. Fonteneuf had seduced or at least encouraged the man to desert, presented his gun at the breast of our leader, but was withheld from firing by the interference of a more sensible comrade. Astonished and angered at the recklessness and audacity of the action, we sprang from our horses, cocked our rifles, and prepared to give them battle, should they presume to offer any further show of hostility. Matters, however, soon assumed a more serious aspect, and we left them to pursue our journey. Immediately after our departure they signified their good will in firing a salute, by the way of bravado, to which, however, we did not think worth while our while to pay any attention. A collision that might have been bloody and fatal, was thus happily, though narrowly, avoided, for in the excitement and passion of the moment, a single shot, fired even by accident, would have been the signal for a deadly encounter. Our progress was necessarily slow from the extreme debility and weakness of our horses, and after marching a few hours, we halted for the night on the margin of a pleasant spring, beneath a grove of cedars, three or four miles west of

Porteneuf. Continuing our route, we reached Snake River, and followed it slowly up to the forks where we opened our spring hunt.

From the mouth of Porteneuf, Snake River flows westward to the falls between forty and fifty miles below, when it gradually turns northward and finally after receiving the waters of several large tributaries, which rise to the westward of the Big Lake, unites itself with Salmon River. It will be recognized on maps of the country as the south fork of Lewis River, but is known among Rocky Mountain hunters by no other name than Snake River, or which is the same thing, "Sho-sho-ne-pah," its ancient appellation, in the language of that Indian tribe. Near the mouth of Porteneuf, it is a broad, magnificent stream, two hundred yards wide, clear, deep, and rapid, bordered by groves of towering cotton wood and aspen trees, and clustering thickets of large willows, matted and bound together by numerous vines and briars. On either side, a vast plain extends its level from thirty to fifty miles in breadth, bounded by ranges of lofty mountains, which in some places are barely visible, owing to their great distance. On the east side of the river, the plain is barren, sandy, and level, and produces only prickly pear, sage, and occasional scanty tufts of dry grass; on the west side the plain is much more extensive than on the opposite, stretching often away to fifty and even sixty miles from the river; it is irregular and sandy, covered with rocks, and like the other, barren of vegetation, except prickly pear and sage - the northern part is a perfect desert of loose, white, sand, dreary, herbless, and arid. It presents everywhere proofs of some mighty convulsion, that sinking mountains to valleys, and elevating valleys to mountains, has changed the aspect of nature, and left the Rocky Mountains in the picturesqueness and grandeur of their present savage and sullen sublimity. Scattered over this immense plain, there are innumerable mounds or masses of rock cracked in the form of a cross at top, (quere, by cooling or the heat of the sun?) the most remarkable of which are three of mountain altitude, situated midway between the mouth of Porteneuf and the mountains northward. Near the largest of these gigantic masses there is a district of some extent covered with huge blocks of black rock, varying in size from a finger stone to a house, which at a distance bear a close resemblance to a village of sombre dwellings.

There are several small rivers flowing from the mountains on this side towards Snake River, not one of which ever reaches it; they are all absorbed by the sand, the strata of which it is evident from this circumstance, must be deep. On the east side, however, there is Porteneuf, and a small river called the Blackfoot, which rises with the sources of Salt River and flows sixty miles westward, to its junction with Snake River, fifteen miles above the mouth of Porteneuf. The Blackfoot is fifty paces in breadth, and is bordered by dense thickets of willow - near the mouth there is a large solitary mound or hill, called the "Blackfoot Butte." Between the Blackfoot and Porteneuf, there is a rich and continuous bottom of excellent grass, where deer are always numerous.

Chapter XV

From the mouth of the Blackfoot, Snake River turns gradually away to the northeast. At the distance of twenty miles, its garniture of grass-covered bottoms and groves of trees entirely disappear, giving place to a more harsh and sterile border, where instead of a rich soil and luxuriant vegetation we find only rocks and sand, with an occasional dwarf cedar, scattering prickley pears, and a gigantic growth of that which flourishes where nothing else can, the everlasting wormwood, or as it is always here called "sage." Twenty miles further up and we come to the junction of "Gray's Creek," a small stream scantily fringed with unneighborly clusters of willows, which rises in "Gray's Hole," between Blackfoot and Salt River, and thence flows north thirty miles and twenty west to its union with Snake River. Fifteen miles above the mouth of Gray's Creek are the "Forks of Snake River," otherwise the junction of "Lewis River" with "Henry's Fork." The first named flows from a cut in the mountain to the southeast, the latter rises with the sources of Madison River in a range of fir covered hills called the "Piny Woods," and runs a southwest course of seventy miles or so to its junction with Lewis River. The country through which it passes after emerging from the Piny Woods is a barren sandy waste, and it is rendered totally unnavigable even for canoes by a succession of falls and rapids. It derives its name from the enterprising Major Henry, who visited the Rocky Mountains shortly after the return of Messrs Lewis and Clark, and built a trading house near its mouth, the remains of which are still visible, bringing sadly to mind the miserable fate of the party left in charge of it, who were overpowered by the Indians and all massacred. It is the "Mad River," mentioned in Chapter V of Coxe's Narrative of adventures on the Columbia River. Lewis River rises with the sources of the Yellow Stone Lake, and flows southward to Jackson's Hole, when it expands to a lake of thirty or forty miles extent, called the "Teton Lake" from a remarkable mountain overlooking it, which bears the name of the "Trois Tetons."
From Jackson's Hole it makes a gradual and graceful sweep to the northwest, until it issues forth from the mountains twenty miles above its mouth, where on the north side a perpendicular wall of rock juts into the plain to a considerable distance, while on the south side its margin is lined by a grove of dwarf cedars that stretches away from the pass or cut several miles. After leaving the mountains it becomes divided into a great number of channels separated from each other by numerous islands, some of which are miles in extent, but others of comparatively small dimensions. Many of these streams or channels are not again united but pursue their several courses till they meet and mingle with the waters of Henry's Fork. There is a high rocky mound in the angle between the two streams, and another on the north side of Henry's Fork, which present the appearance of having been once united. Both these mounds are large and lofty and may be easily seen from the plain at a distance of from thirty to forty miles. There are likewise two or three similar but smaller mounds on an island in Lewis

River, their summits just appearing above the forest of cottonwood trees by which they are entirely surrounded and nearly concealed. Noble groves of aspen and cottonwood, and dense thickets of willow border all these streams and channels, and form almost impenetrable barriers around the verdant prairies covered with fine grass and delicate rushes which lie embowered within their islands. Deer and elk in great numbers resort to these fair fields of greenness in the season of growth, and during the inclemencies of winter seek shelter in the thickets by which they are environed, where rushes of gigantic size, exceeding in stature the height of man, are found in wild profusion. Flowing into Henry's Fork there is, a short distance above Lewis River, a small stream called Pierres Fork. It rises in Pierre's Hole, and has a westerly course of sixty miles to its mouth. Twenty miles above the forks of Snake River, Henry's Fork is subdivided into two streams of nearly equal magnitude, one of which before leaving the mountains bounds over a lofty precipice, thus forming a most magnificent cascade. Lewis River is about two hundred miles long, and receives in its course several streams which I shall hereafter have occasion to notice.

During our journey to the forks of Snake River, we saw and killed numbers of Buffalo, and saw also hundreds of their carcases floating down the river, or lodged with drift wood upon the shoals. These animals were probably drowned by breaking through when endeavoring to cross the river on the ice. At Gray's Creek we came in view of the "Trois Tetons," (three breasts) which are three inaccessible finger-shaped peaks of a lofty mountain overlooking the country to a vast distance. They were about seventy miles to the northeast, when observed from that point. Their appearing is quite singular, and they form a noted land mark in that region. During our march we had several hard showers of rain, and occasionally a storm of sleet, but the weather was generally mild and pleasant. From the Forks of Snake River we continued up to the forks of Henry's Fork, trapping as we went and taking from forty to seventy beaver a day. Some of them were large and fat, and when well boiled proved to be excellent eating. Our cuisine was not of the best perhaps, but we made up in plenty what we lacked in variety, and on the whole fared very tolerably. As we ascended Henry's Fork, trees and grass again disappeared, and the waters of both branches were frequently compressed to narrow channels by bold bluff ledges of black rock, through which they darted with wild rapidity, and thundering sound.

A small party of hunters was sent to the "Burnt Hole," on the Madison River, in quest of beaver, but returned back without success. The Burnt Hole is a district on the north side of the Piny woods, which was observed to be wrapped in flames a few years since. The conflagration that occasioned this name must have been of great extent, and large forests of half consumed pines still evidence the ravages at that time of the destructive element. The Piny Woods are seen stretching darkly along like a belt of twilight from south of east to northwest, distant from us about thirty miles.

From Henry's Fork we passed westward to the head of "Kammas Creek," (so called from a small root, very nutritious, and much prized as food by Indians and others, which abounds here,) a small stream that rises with the sources of the

Madison, flows southeast forty or fifty miles, and discharges itself into a pond six miles northwest from the mouth of Gray's Creek. This pond has no outlet, its surplus waters pass off by evaporation, or are absorbed by the sand. During our stay on this stream, several Indians were seen lurking about, and evidently watching an opportunity to steal our horses, or commit some other depredation upon our party.

On the twenty-eighth of May, two of our men, Daniel Y. Richards, and Henry Duhern by name, went out as usual to set their traps, but never returned. We ascertained subsequently, that they were butchered by a war party of Blackfeet Indians. Four of our men, crossing over to the southeastern sources of the Jefferson, discovered a number of mounted Indians, and fled back to camp in alarm. At the same time, a party of Flathead Indians came to us from a village four days' march to the northwestward. They informed us that they had a skirmish a few days previous with a party of Blackfeet, two of whom they killed. Several of the Flathead warriors were immediately dispatched to their village with a present to the chief, and a request that he and his people would come and trade with us. During their absence, we moved westward to a small stream called "Poison Weed Creek," from a deleterious plant found in its vicinity. The waters of this creek also, are drank up by the thirsty sands. Large herds of buffalo were driven over to us before the Flatheads, many of which we killed, and about one out of a dozen of which was found fat enough to be palatable. Several of the young Flatheads in the mean time joined us in advance of the village. The day previous to their coming, one of their tribe was killed by the Blackfeet, who also caught a Flathead squaw some distance from the village, whom they treated with great barbarity - they ravished her, cut off her hair, and in this condition sent her home.

Two or three days after their arrival, the whole village, consisting of fifty lodges of Flatheads, Nezperces, and Pend'orielles, came in sight, but unlike all other Indians we have hitherto seen, they advanced to meet us in a slow and orderly manner singing their songs of peace. When they had approached within fifty paces, they discharged their guns in the air, reloaded, and fired them off again in like manner. The salute of course, was returned by our party. The Indians now dismounted, left their arms and horses, and silently advanced in the following order: first came the principle chief, bearing a common English flag, then four subordinate chiefs, then a long line of warriors, then young men and boys who had not yet distinguished themselves in battle, and lastly the women and children, who closed the procession. When the Chief had come up, he grasped the hand of our Partizan, (leader,) raised it as high as his head, and held it in that position while he muttered a prayer of two minutes duration. In the same manner he paid his respects to each of our party, with a prayer of a minute's length. His example was followed by the rest, in the order of rank. The whole ceremony occupied about two hours, at the end of which time each of us had shaken hands with them all. Pipes were then produced, and they seated themselves in a circle on the ground, to hold a council with our leaders respecting trade.

The Flatheads probably derive their name from an ancient practice of shaping or deforming the head during infancy, by compressing it between boards placed on the forehead and back part, though not one living proof of the existence at any time of that practice can now be found among them. They call themselves in their beautiful tongue, "Salish," and speak a language remarkable for its sweetness and simplicity. They are noted for humanity, courage, prudence, candour, forbearance, integrity, trustfulness, piety, and honesty. They are the only tribe in the Rocky Mountains that can with truth boast of the fact that they have never killed or robbed a white man, nor stolen a single horse, how great soever the necessity and the temptation. I have, since the time mentioned here, been often employed in trading and travelling with them, and have never known one to steal so much as an awl-blade. Every other tribe in the Rocky Mountains hold theft rather in the light of a virtue than a fault, and many even pride themselves on their dexterity and address in the art of appropriation, like the Greeks deeming it no dishonour to steal, but a disgrace to be detected.

Chapter XVI

The Flatheads have received some notions of religion either from pious traders or from transient ministers who have visited the Columbia. Their ancient superstitions have given place to the more enlightened views of the christian faith, and they seem to have become deeply and profitably impressed with the great truths of the gospel. They appear to be very devout and orderly, and never eat, drink, or sleep, without giving thanks to God. The doctrines they have received are no doubt essential to their happiness and safety in a future state of existence, but they oppose, and almost fatally, their security and increase in this world. They have been taught never to fight except in self defence, or as they express it, "never to go out to hunt their own graves," but to remain at home and defend manfully their wives and children when attacked. This policy is the worst that could be adopted, and is indeed an error of fatal magnitude, for the consequence is that a numerous, well armed, watchful, and merciless enemy, with whom they have been at war from time immemorial, emboldened by their forbearance, and puffed up with pride by their own immunity, seek every occasion to harass and destroy them, - steal their horses, butcher their best hunters, and cut them off in detail. Fearing to offend the Deity, they dare not go out to revenge their murdered friends and kinsmen, and thus inspire their blood-thirsty foes with a salutary dread of retributive justice; and hence they are incessantly exposed to the shafts of their vindictive enemies, outlying parties of whom are almost constantly on the watch to surprise and massacre stragglers, unrestrained by the fear of pursuit and vengeance. Under the influence of such an untoward state of things, they are rapidly wasting away, in spite of courage, patriotism, and many virtues that have no parallel in the Rocky Mountains.

Though they defend themselves with a bravery, skill, and devotion that has absolutely no comparison, proving on every occasion their great superiority in dauntlessness and address, no advantages of daring and prowess can overcome the evil effects of their defensive policy, and the probability is that in a few years more the noblest race of uncivilized men, will become utterly extinct.

Many anecdotes of Messrs. Lewis and Clark, who were the first white men they ever saw, are related by the Flatheads, and some of the old men in the village now with us, were present at their first interview. An intelligent Flathead, known to the hunters by the name of "Faro," related to me many curious incidents in their history, and among others an account of this first interview with the whites, which, though obtained two years later in point of time, may not be uninteresting in this connexion. I give it nearly in his own language.

"A great many snows past," said he, "when I was a child, our people were in continual fear of the Blackfeet, who were already in possession of fire arms of which we knew nothing, save by their murderous effects. During our excursions for buffalo, we were frequently attacked by them, and many of our bravest warriors fell victims to the thunder and lightning they wielded, which we conjectured had been given them by the Great Spirit to punish us for our sins. In our numerous conflicts, they never came in reach of our arrows, but remained at such a distance that they could deal death to us without endangering themselves. Sometimes indeed their young warriors closed in with us, and were as often vanquished; but they never failed to repay us fourfold from a safe distance. For several moons we saw our best warriors almost daily falling around us, without our being able to avenge their deaths. Goaded by thirst for revenge, we often rushed forth upon our enemies, but they receded like the rainbow in proportion as we advanced, and ever remained at the same distance, whence they destroyed us by their deadly bolts, while we were utterly powerless to oppose them. At length, 'Big Foot,' the great chief of our tribe, assembled his warriors in council, and made a speech to them, in which he set forth the necessity our leaving our country. 'My heart tells me,' said he, 'that the Great Spirit has forsaken us; he has furnished our enemies with his thunder to destroy us, yet something whispers to me, that we may fly to the mountains and avoid a fate, which, if we remain here is inevitable. The lips of our women are white with dread, there are no smiles on the lips of our children. Our joyous sports are no more, glad tales are gone from the evening fires of our lodges. I see no face but is sad, silent, and thoughtful; nothing meets my ears but wild lamentations for departed heroes. Arise, let us fly to the mountains, let us seek their deepest recesses where unknown to our destroyers, we may hunt the deer and the bighorn, and bring gladness back to the hearts of our wives and our children!'

"The sun arose on the following morning to shine upon a deserted camp, for the little band of Flatheads were already leaving the beautiful plains of the Jefferson. During one whole moon we pursued our course southwestward, through devious paths and unexplored defiles, until at last, heartsore and weary, we reached the margin of salmon river. Here we pitched our camp, and whilst the women were employed in gathering fruits and berries, our hunters explored the surrounding mountains, which they found stored with abundance of game, as the stooping trees and bushes that grew around our lodges, told us on our return; we likewise made the joyful discovery that the river was alive with salmon, great numbers of which were taken and preserved against future necessity. The Great Spirit seemed again to look kindly upon us. We were no longer disturbed by our enemies, and joy and gladness came back to our bosoms. Smiles like little birds came and lit upon the lips of our children, their merry laughter was a constant song, like the song of birds. The eyes of our maidens were again like the twinkling stars, and their voices soft as the voice of a vanishing echo. There was plenty in every lodge, there was content in every heart. Our former pastimes were renewed, our former fears were forgotten. Pleasant tales again wooed the twilight, and the moon was the only watch that we kept upon our slumbers. Our hunters went out in safety, there was no blood upon the path. They came back loaded with game, there was no one to frighten away the deer. Peace hovered around our council fires, we smoked the calumet in peace.

"After several moons, however, this state of tranquil happiness was interrupted by the unexpected arrival of two strangers. They were unlike any people we had hitherto seen, fairer than ourselves, and clothed with skins unknown to us. They seemed to be descended from the regions of the great "Edle-a-ma-hum." They gave us things like solid water, which were sometimes brilliant as the sun, and which sometimes showed us our own faces. Nothing could equal our wonder and delight. We thought them the children of the Great Spirit. But we were destined to be again overwhelmed with fear, for we soon discovered that they were in possession of the identical thunder and lightning that had proved in the hands of our foes so fatal to our happiness. We also understood that they had come by the way of Beaver-head River, and that a party of beings like themselves were but a day's march behind them.

"Many of our people were now exceedingly terrified, making no doubt but that they were leagued with our enemies the Blackfeet, and coming jointly to destroy us. This opinion was strengthened by a request they made for us to go and meet their friends. At first this was denied, but a speech from our beloved chief, who convinced us that it was best to conciliate if possible the favor of a people so terribly armed, and who might protect us, especially since our retreat was discovered, induced most of our warriors to follow him and accompany the strangers to their camp. As they disappeared over a hill in the neighborhood of our village, the women set up a doleful yell, which was equivalent to bidding them farewell forever, and which did any thing but elevate their drooping spirits.

"After such dismal forebodings imagine how agreeably they were disappointed, when, upon arriving at the strangers encampment, they found, instead of an overwhelming force of their enemies, a few strangers like the two already with them, who treated them with great kindness, and gave them many things that had not existed before even in their dreams or imaginations. Our eagle-eyed chief discovered from the carelessness of the strangers with regard to their things, that they were unacquainted with theft, which induced him to caution his followers against pilfering any article whatever. His instructions were strictly obeyed, mutual confidence was thus established. The strangers accompanied him back to the village, and there was peace and joy in the lodges of our people. They remained with us several days, and the Flatheads have been ever since the friends of the white men."

Chapter XVII

Gambling seems not to be disallowed by the religion of the Flatheads, or rather perhaps is not included among the number of deadly offences, for they remain incurably addicted to the vice, and often play during the whole night. Instances of individuals losing everything they possess are by no means infrequent. Their favourite game is called "Hand," by the hunters, and is played by four persons or more. - Betters, provided with small sticks, beat time to a song in which they all join. The players and betters seat themselves opposite to their antagonists, and the game is opened by two players, one of each side, who are provided each with two small bones, one called the true, and the other false. These bones they shift from hand to hand, for a few moments with great dexterity, and then hold their closed hands, stretched apart, for their respective opponents to guess in which the true bone is concealed. This they signify by pointing with the finger. Should one of them chance to guess aright and the other wrong, the first is entitled to both true bones, and to one point in the game. Points are marked by twenty small sharp sticks, which are stuck into the ground and paid back and forth until one side wins them all, which concludes the game. The lucky player, who has obtained both the true bones, immediately gives one to a comrade, and all the players on his side join in a song, while the bones are concealing. Should the guesser on the opposite side miss both the true bones, he pays two points, and tries again; should he miss only one, he pays one point. When he guesses them both, he commences singing and hiding the bones, and so the game continues until one, or other of the parties wins. They have likewise a game called by the French name of "Roulette." This game is played by two persons with a small iron ring, two or three inches in diameter, having beads of various colours fastened to the inside. The ring is rolled over a piece of smooth ground by one of the players; both follow it and endeavour to pitch arrows so that the ring may fall upon them. The beads are of different values, and such only count as may happen to be directly over the arrow. The points, of the game are counted by small sticks, and the winning of a stated number determines it. Throwing arrows at a target with the thumb and finger, is a common game with the boys; shooting at a mark is also much practiced.

The women are as much addicted to gaming as the men. They play at Hand, and have also a game which is never played by the other sex. Four bones eight inches in length, which are marked on one side with figures common to two of them, are thrown forward on a buffalo robe, spread down for the purpose. If the white sides of all the four fall uppermost, they count four, and throw again; if two figured ones of the same kind, and two white ones are up, they count two, but if one odd one should be turned, they count nothing, and the adverse party takes the bones with the same privilege. This game is won by the party which gets an expressed number of points first.

Horse racing is a favourite amusement, with the Flatheads. In short races they pay no attention to the start, but decide in favour of the horse that comes out foremost. Sometimes in long races they have no particular distance assigned, but the leading horse is privileged to go where he pleases, and the other is obliged to follow until he can pass and take the lead. These races generally terminate in favour of bottom rather than speed. Occasionally they have club races, when they enter such horses, and as many of them as they please, run to some certain point and back, when the foremost horse is entitled to all the bets. These games and races are not peculiar to the Flatheads, but are common to all the tribes we have met with in the country.

We remained on Poison-weed creek with the Flatheads until the 19th of June, when Messrs. Fontenelle and Dripps, with thirty men departed for St. Louis. They were accompanied by twenty Flatheads to Cache Valley, where they expected to meet the Rocky Mountain Fur Company, under a mutual agreement to return together to St. Louis. The remainder of our party, with the Iroquois amounting to twenty-five persons, set out at the same time for Salmon River, in company with the Flatheads. A Snake Indian came into our camp on the 22d, and informed us that he was one of two hunters, who escaped from a general massacre of the inmates of six lodges on Salmon river three days previous. Himself and companion returning from a hunting expedition, instead of the friends and relatives they expected to meet them and welcome them home from the fatigues of the chase, found only scalpless bodies and desolate lodges. Old and young, weak and strong, homely and fair, had met a common doom; not one was left alive of all they had so lately parted from in health and happiness. It was a sad tale, yet no uncommon one in this region of barbarism and inhumanity. The next day four of the Flatheads who went with Fontenelle, returned and reported that he had a slight skirmish on Snake river with a party of Blackfeet, of whom they killed one, and took five horses, his own party sustaining no loss or injury.

On the 28th day of June we ascended a creek that flows from Day's defile, and unites with another from the north called Cotas Creek, both falling into a pond situated at the eastern extremity of a point of the mountain jutting down in the plain on the south side of Day's Creek. Day's defile receives its name from John Day, a noted hunter, who died and was buried here a few years since. We journeyed slowly, being engaged in procuring and drying meat for our subsistence during the fall hunt, which it is intended to make in a rugged country on the sources of Salmon river, where white men had never penetrated, but where beaver was said to be abundant. A Snake Indian, engaged for a guide, states that buffalo are no where to be found in that wild district, and hence the necessity of securing a good supply of provisions to be taken with us.

From the outlet of Day's Creek, we proceeded up forty miles to its source, and thence continued over a narrow pass between two mountains which we found so free from all obstructions that even wagons might cross with ease, and which conducted us to a large valley watered by a small stream called Little Salmon River, which flows through it in a western course. The source of this stream, though only ten or twelve feet wide, was yet so deep as to be unfordable, except at occasional points. Some of our men, ignorant of its depth, attempted to ford it, but only escaped drowning, by clinging to the branches which were interlaced and bound together by wild vines, forming a complete canopy over the stream. Their horses were carried some distance down by the impetuosity of the current before we could reach and rescue them.

We passed through a defile, on the first of July bearing southward, which was dotted occasionally, with banks of snow, which were however rapidly disappearing. This defile brought us to a small valley watered by Gordiaz river, which, rising with the sources of the Malade, against those of Salmon river, flows eighty to one hundred miles eastward, gradually decreasing in breadth and depth, until it finally disappears in the plains of Snake river twenty-five miles north of the mouth of the Blackfoot. We found this valley covered with Buffalo, many of which we killed; we remained here until the 5th to dry the meat.

We now separated from the Flatheads, (except a few lodges, who remained with us,) and crossed a mountain to the westward, when we reached a torrent bounding over the rocks in that direction, which we followed for several miles, until it falls into a large one from the southwestward. This we ascended to its source, a small lake, one or two miles in extent, of great depth and perfectly transparent, situated in a hollow corner or cavity on the summit of a mountain. From hence we travelled westward over a harsh rugged pine covered country, destitute of valleys, but abounding with deep ravines, and dark gulfs. The sides of the mountains were often so abrupt that our horses were continually falling down, and frequently fell fifty or sixty feet before they could recover their footing. In some places the fallen pines were so numerous, and withal so interlocked with each other, that we were forced to cut a passage through. The northern declivities of all these mountains were covered with snow banks, or chaotick masses of snow and rock intermingled, which had fallen from the summits. All the streams that wind through the deep cavities or gulfs between these mountains, are roaring and tumbling torrents, which flow eastward, and fall into Salmon river. Over these mountains, gulfs, ravines and torrents, and a thousand other obstacles we pursued our difficult and toilsome way, often at the imminent hazard of our lives, and compelled often to retrace our weary steps; sometimes journeying above the clouds, and again through passages so deep and dark, that no straggling sunbeam ever pierced their gloom. Thus alternated our course over mountain heights, and through tartarian depths.

Chapter XVIII

After a tedious and toilsome march, we at length encamped on the 13th in a prairie, forming the central portion of a large valley half grown up with lofty pines, which is watered by one of the largest and most westerly of the sources of Salmon river. Here we found a party of "Root Diggers," or Snake Indians without horses. They subsist upon the flesh of elk, deer, and bighorns, and upon salmon which ascend to the fountain sources of this river, and are here taken in great numbers. These they first split and dry, and then pulverize for winter's provision. They often, when unable to procure fish or game, collect large quantities of roots for food, whence their name. We found them extremely anxious to exchange salmon for buffalo meat, of which they are very fond, and which they never procure in this country, unless by purchase of their friends who occasionally come from the plains to trade with them. We have not seen a vestige of buffalo since leaving the valley of Gordiez river.

From observing that many of these Indians were clad with robes and moccasins made of dressed beaver skins, we were induced to believe that the information we had previously received in regard to the abundance of those animals in this vicinity was true. Our enterprising hunters forthwith engaged one of these Indians to serve as a guide, and set out on a trapping expedition, not doubting but that they should return in a few days with horse loads of fur. Meanwhile such of us as remained to take care of camp, were employed in taking salmon, which was easily effected by driving them up or down the river, over shoals and rapids where we killed them with clubs and stones, and frequently even caught them with our hands.

Our horses were daily so much annoyed by flies, that they were forced to assemble in crowds for their mutual defence, and were seen switching and brushing one another continually with their tails in the most affectionate and friendly manner. Hence I infer that among animals, of an inferior order to man at least - the strongest bands of friendship are forged more by interest than inclination. Our poor beasts from having nothing else to rub against, in the open prairie, were compelled to rub against each other to get rid of their tormenters, and thus necessity forced them to mutual kind offices, and established among them a community of friendly feelings and acts of generosity. Are the ties of social and political union among men often of a more refined and liberal character than that which bound together these poor, fretted animals in an intercourse of mutual amelioration?

At the expiration of ten days our hunters returned with ill success impressed most audibly upon their downcast visages. They reported that their guide conducted them about fifty miles further west and showed them a small group of beaver lodges, from which they caught some thirty of those animals. This accomplished they desired the Indian to proceed. He then led them to the summit of a lofty mountain, overlooking a vast plain watered by several streams, whose borders were garnished with groves of aspen and cottonwood trees, and pointing down with his finger inquired, "Do you see those rivers?" "Yes," returned the trappers, "but are there any beaver there?" "No," answered the Indian, with animation, "but there is abundance of elk." In the first heat of their indignation they could scarcely refrain from killing the poor Indian, who beheld with astonishment their anger at the receipt of information, which in his simplicity he had supposed must give them great delight. A moment's reflection, however, satisfied them that their guide meant well, though he had deceived them sorely. The truth was that the Indian imagining they hunted merely for food, had prepared them what he thought would be a most agreeable surprise, in leading them to where instead of the humble beaver they would find the lordly elk. There was nothing left for them to do but to return to camp, which they did with all expedition, leaving the wandering guide gazing alternately at the inviting prospect spread before him, and at the retreating cavalcade of retiring trappers and vainly striving to read the riddle of their disappointment and departure. Having thus satisfied ourselves that our visit to this now interesting country was a complete failure, we determined to retrace our steps with the utmost possible despatch to the plains, and make our hunt elsewhere. Accordingly on the 24th we commenced our return travel, (I had almost spelt it with an ai, instead of an e, more, however, for the sake of the truth than a bad pun;) and for some time wandered about in almost every direction, to avoid the numerous obstacles that impeded nearly every step of the way, though our general course was towards the rising sun.

During our journey, I witnessed the process of cooking "Kamas," a small root about the size of a crab apple, which abounds in many parts of this country, in the rich bottoms that border most of the streams and rivers. The mode of preparing this root, is almost identical with that by which the south sea Islanders cook their cannibal and swinish food, and the west Indians their plantain. The squaws, by whom all the avocations of domestic labour are performed, excavate round holes in the earth two feet deep, and three in diameter, which are then filled with dry wood and stones in alternate layers, and the fuel fired beneath. When the wood consumes the heated stones fall to the bottom, and are then covered with a layer of grass, upon which two or three bushels of kamas roots, according to the capacity of the whole, are placed, and covered with a layer of grass, and the whole coated over with earth, upon which a large fire is kept burning for fifteen hours. Time is then allowed for the kamas to cool, when the hole is opened, and if perfectly done, the roots which were before white, are now of a deep black colour, not disagreeable to the taste, and having something the flavour of liquorice. Thus prepared, the kamas is both edible and nutritious, and forms no inconsiderable item of food with many of the Rocky Mountain tribes.

We found ourselves in the beginning of August, much to our satisfaction again in a level country, and far from the sombre folds of the mountain-wrapping pine in the dense forests of which the brightest day is but a starless twilight, and the fairest evening but a thick and blackened night. We rested from our mountain toils - toils in a double sense - in a beautiful valley twelve miles long, and from four to five broad, intersected by several willow and aspen bordered streams, tributary to Salmon River, which flows through the valley in a northeast direction. The river is here one hundred yards wide, clear, shallow, and arrowy swift. We had a shower of rain on the fourth, the first that has fallen since the middle of June.

On the eleventh, we fell in with the Flatheads, from whom we had parted a month before. Nothing worthy of record had chanced among them since our separation. In the afternoon, two horsemen were observed on a neighbouring bluff, but concealed themselves or fled ere they could be reached by a party of Flathead warriors, who were speedily mounted and in pursuit.

After this period we returned by way of the valley, of the Gordiez river, and little Salmon river to Day's Defile. During our route we saw traces of footmen, and one evening, heard the reports of firearms from a neighboring mountain, but saw no strange Indians, and met with no disaster of consequence. We killed several grizzly bears and a variety of other game. From the head of Day's creek, we crossed a mountain eastward to "Cota's Defile", so named from a man who was shot while performing a sentinel's duty, one dark night, by an Indian. On the 19th we had a snow storm of several hours duration. In the valleys the snow melted as fast as it fell, but the surrounding mountains were whitened with it for two days.

Cota's Defile brought us to the head waters of the east fork of the Salmon river, in an extensive valley, thirty miles long and ten to twelve in breadth. The principal stream is forty paces wide, bordered with willows, and birch and aspen, and flows northwestward fifty miles to Salmon river. From the summit of Cota's defile we saw a dense cloud of smoke rising from the plains forty or fifty miles to the southeastward, which we supposed to have been raised by the Flatheads, who accompanied Fontenelle to Cache Valley, and who were now in quest of the village to which they belong. The Indians with us answered the signal by firing a quantity of fallen pines on the summit of a high mountain.

It may seem to the reader a trifling matter to note the track of footmen, the report of firearms, the appearance of strange horsemen, and the curling vapour of a far off fire, but these are far from trivial incidents in a region of country where the most important events are indiced by such signs only. Every man carries here emphatically his life in his hand, and it is only by the most watchful precaution, grounded upon and guided by the observation of every unnatural appearance however slight, that he can hope to preserve it. The footmark may indicate the vicinity of a war party hovering to destroy; the report of firearms may betray the dangerous neighbourhood of a numerous, well armed, and wily enemy; strange horsemen may be but the outriding scouts of a predatory band at hand and in force to attack; the rising smoke may indeed curl up from the camp of friends or an accidental fire, but it more probably signals the gathering forces of an enemy recruiting their scattered bands for the work of plunder and massacre. Thus every strange appearance becomes an important indication which the ripest wisdom and experience are needful to interpret; and the most studious care and profound sagacity are requisite to make the most advantage from. It is only in this manner that the hunter's life is rendered even comparatively secure, and it is thus that the most trivial occurrence assumes a character of the gravest moment, freighted as it may be with the most alarming and perilous consequences.

Chapter XIX

On the 25th of August we again separated from the Flatheads, except a few lodges which accompanied us wherever we went, and entering a narrow cut in the mountain on the east side of the valley, followed the stream that flows through it to its source, and thence crossing a prairie hill descended into Horse Prairie fifteen miles north of the East Fork valley. Horse Prairie is a pleasant rolling plain fifty or sixty miles in circumference and surrounded by lofty mountains. In the middle of the valley there is a conical rocky mound rising from the plain on the north side of the stream, and directly fronting a high rocky bluff on the opposite side. These elevations, separated by a few hundred feet, at a distance convey the idea of a formidable gateway. The borders of the creeks and rivulets in this valley are scantily adorned with clusters of small willows. The largest of these creeks flows through the valley to the northeastward, and is the most southwesterly source of the Jefferson. The Sho- sho-ne Cove, where Capt. Lewis in advance of the canoes and with one attendant, discovered the first Rocky Mountain Indian whose confidence he endeavored to win by friendly signs and the offering of trinkets, but who, timid as a hare, fled in the mountains westward and crossed the Salmon River, is in this valley.

A trapper of our party by the name of Perkins was fired upon on the 27th from a thicket near which he happened to be passing, but fortunately escaped uninjured, though the ball passed through the left breast of his coat. His horse, alarmed by the sudden report of the gun, sprang forward throwing off by the action his fusil which he was carrying carelessly across his saddle, seeing this an Indian sprang forth instantly from the thicket and pounced upon it, but before he could bring it to bear upon the trapper the latter by dint of whip and spur and a fleet steed had contrived to get beyond his reach. A party of hunters returned with Perkins to the scene of his discomfiture, but the Indians had already taken their departure, and that with such precipitation that several trifling articles were overlooked in their haste and left behind them.

We left Horse Prairie on the last day of the month, and crossing the mountain northwestward, descended into the Big Hole. This is an extensive valley of sixty miles length, and fifteen to twenty broad, bounded on every side by lofty, irregular and picturesque ranges of mountains, the bases of which are girded with dense forests of fir which in some places encroach upon the prairie domain. Above the pine region, the mountains present immense pointed masses of naked rock, hiding their giant heads among the clouds where the eye vainly strives to follow; and often even piercing through the misty realm, where storm spirits hold their frolic revels, so that their gray peaks are often seen flashing and basking in the sun while the thickening vapours below are sending down torrents of rain, and it may be belting their hoary forms with lightning lines of fire, and beating their stolid breasts with blows and bolts of thunder, or darkening the atmosphere with heavy falls of snow and hail. The caverns or gulfs - they are not vales - between these worlds of rock are heaped with the snows of ages.

This valley is watered by innumerable willow-fringed streams that unite and form Wisdom River, which flows a little east of north, and, after leaving the valley, eastward, to its junction with the Jefferson distant eighty miles.

On the first of September I discovered the burrow of a species of beautiful small spotted fox, and wishing to obtain one of their skins, sent an Indian boy to camp for a brand of fire designing, if possible, to drive them out by the aid of smoke. The careless boy scattered a few sparks in the prairie, which, the dry grass almost instantly igniting, was soon wrapped in a mantle of flame. A light breeze from the south carried it with rapidity down the valley, sweeping everything before it, and filling the air with black clouds of smoke. Our absent trappers returned at full speed, expecting to find camp attacked or at least the horses stolen, but were agreeably disappointed on learning the real nature of the accident. It however occasioned us no inconsiderable degree of uneasiness as we were now on the borders of the Blackfoot country, and had frequently seen traces of small parties, who it was reasonably inferred might be collected by the smoke, which is their accustomed rallying signal, in sufficient force to attack us. Our party consisted of thirty armed men, a mere handful when compared to the prairie-reddening parties of Blackfeet which are often seen here. Clouds of smoke were observed on the following day curling up from the summit of a mountain jutting into the east side of the valley, probably raised by the Blackfeet to gather their scattered bands, though the truth was never more clearly ascertained.

We were detained on the tenth, by a storm of snow which covered the earth to the depth of several inches, but disappeared on the following night. During our stay at this encampment we found the petrified trunk of a large cedar half imbedded in the earth. Next day we left the Big Hole by its northern extremity and crossed a mountain to the Deer Horse Plains. This is a valley somewhat larger than the Big Hole, and like that surrounded by mountains, generally however low, barren and naked, except to the south and east where lofty and snowclad peaks appear. All the streams by which it is intersected are decorated with groves and thickets of aspen, birch and willow, and occasional clusters of currant and gooseberry bushes. The bottoms are rich and verdant, and are resorted to by great numbers of deer and elk. The several streams unite and form "La Riviere des pierres a fleches," (Arrow Stone River,) thus named from a kind of semi-transparent stone found near it, formerly much used by the Indians for making points of arrows. This river is one of the sources of Clark's River, and flows through the valley to the northeastward. The valley owes its singular but appropriate name to a natural curiosity situated near the river a few miles from the eastern side. The curiosity referred to is a semi-spherical mound some fifty paces in circumference and fifteen feet high, rather flattened at top, and covered with turf and a sickly growth of yellow grass. There are several cavities in the highest part of the mound, the largest measuring a foot in diameter, in all of which water is seen boiling a few inches below the surface. The earth is heated but not to such a degree as to prevent vegetation, except about immediate edges of the cavities. This mound, like those on Snake River, has been evidently self-formed by continual deposits of calcareous cement, hardened to the consistence of rock. How the soil came upon its summit is matter of inquiry, perhaps by the encroachments and decay of the creeping vegetation of years. The ground about its base is low and marshy, and several transparent pools of tepid water near by, are famous resorts for bathing by the Indians. These waters are slightly impregnated with salt, which quality renders the place attractive to deer, and it is seldom without visiters of this description. Animals as well as men have their favourite (not to say fashionable,) watering places, and this is one of them. Clouds of vapour are continually emanating from the mound, which at a distance on a clear cold morning might readily be mistaken for smoke, - the mound itself has much the resemblance of an Indian Cabin, and hence which the name by the valley is designated. The water within the mound is so hot one cannot bear a finger in it for a moment. The presence of sulphur is shown by the unmistakable, and any thing but fragrant smell of the vapour.

On entering the Deer House Plains we were alarmed by the cry of Indians from the advance guard of the party, but almost as quickly freed from apprehension by the arrival of a Pen-d'orielle, who gave us to understand, that one hundred lodges of his tribe lay encamped eight miles below. - Early next day they removed their quarters and took up a position in the immediate vicinity of our own, when we ascertained that they were on their way from the Flathead Trading House of the Hudson Bay Company to buffalo, and were living upon a mixed diet of roots and expectations, the latter in much the larger proportion - plainly they were nearly starving. It is a well ascertained fact, that buffalo confine their range to the eastern side of an imaginary line commencing at the south on the west side of the Arkansas, in about Latitude 38 north and Longitude 28 west from Washington, and running thence around the headwaters of the Arkansas, crossing the sources of the Rio Grande, Blue River (its principal branch) and Salt River, then turning from north to west in a nearly direct course - crossing Green River above the mouth of Ashley's Fork, - to the Big Lake at the mouth of Bear River, thence crossing Salt River a short distance below the junction of Porte Neuf, and through the sources of Salmon River to a point in Latitude 44E40' north, and Longitude 33E20' west or nearly, thence around the sources of Clark's River on the east, and thence in a north west direction west of the Missouri, and its principal sources to Lat. 49 north and Longitude 34E20' west, where it passes the limit of my observation and inquiries to the northward. East of this line they range back and forth across the great plains of the Mississippi and Missouri, retiring towards the mountains in the winter, and in spring spreading themselves over the vast prairies, and almost blackening the waste by their countless numbers, from the Pawnee hunting ground, to the far off ranges of the Rocky Mountains.

As the line described almost skirts the Deer House Plains buffalo are seldom found west of this valley, and rarely even here, which was now the case. Indeed we have seen none since leaving the east fork of the Salmon River, though Horse Prairie is a famous resort for them, and they sometimes penetrate to the Big Hole.

We were annoyed almost beyond endurance by the hundreds of famishing dogs belonging to the Indians. They devoured every leathern article that lay within reach, even to the bull-hide thongs, with which we fastened our horses. We were compelled to keep guard by turns or risk the entire loss of our baggage, their depredations were so bold and incessant. I performed my watch at the salient angle of our tent, armed with an axe, which I hurled among them without respect to "mongrel, puppy, whelp or hound," and not infrequently sent some of them back yelping a serenade of pain to their sleeping masters. Once, however, on returning with the axe which I had thrown unusually far, I discovered a scury cur, coolly trotting off with my saddle bags, which the rascal had stolen from within the protection of the tent. It is needless to say that I pursued and recovered them, but ere I could return to my post, I perceived three large fellows marching leisurely homeward, with a bale of dried meat, weighing not less than forty pounds. Grounding an inference hereupon that in spite of the axe and my utmost efforts they would prove victorious, I thought it advisable to let my manhood take care of itself, and call up my dreaming companions. No sooner said than done, when we called a council of war, and deeming discretion with such an enemy the better part of valour, we suspended all our baggage in a tree that overhung the tent, and went to rest without apprehension of the consequences. In the morning we found every thing safe as we had left it, while our less careful neighbours were seen busily collecting the scattered relics of the night's devastation. One of them lost above forty dollars' worth of furs, and another, a jolly old Frenchman, drew his pipe from his teeth to swear with more emphasis that the scoundrelly dogs had devoured his axe.

Chapter XX

We departed southeastward for the Jefferson River on the morning of the fifteenth, accompanied by all the Indians; and picturesque enough was the order and appearance of our march. Fancy to yourself, reader, three thousand horses of every variety of size and colour, with trappings almost as varied as their appearance, either packed or ridden by a thousand souls from squalling infancy to decrepid age, their persons fantastically ornamented with scarlet coats, blankets of all colours, buffalo robes painted with hideous little figures, resembling grasshoppers quite as much as men for which they were intended, and sheep-skin dresses garnished with porcupine quills, beads, hawk bells, and human hair. Imagine this motley collection of human figures, crowned with long black locks gently waving in the wind, their faces painted with vermillion, and yellow ochre. Listen to the rattle of numberless lodgepoles trained by packhorses, to the various noises of children screaming, women scolding, and dogs howling. Observe occasional frightened horses running away and scattering their lading over the prairie. See here and there groups of Indian boys dashing about at full speed, sporting over the plain, or quietly listening to traditional tales of battles and surprises, recounted by their elder companions. Yonder see a hundred horsemen pursuing a herd of antelopes, which sport and wind before them conscious of superior fleetness,- there as many others racing towards a distant mound, wild with emulation and excitement, and in every direction crowds of hungry dogs chasing and worrying timid rabbits, and other small animals. Imagine these scenes, with all their bustle, vociferation and confusion, lighted by the flashes of hundreds of gleaming gun-barrels, upon which the rays of a fervent sun are playing, a beautiful level prairie, with dark blue snow-capped mountains in the distance for the locale, and you will have a faint idea of the character and aspect of our march, as we followed old Guignon (French for bad-luck) the Flathead or rather the Pen-d'oreille chief slowly over the plains, on the sources of Clark's River. Exhibitions of this description are so common to the country that they scarcely elicit a passing remark, except from some comparative stranger.

Next day we separated into two parties, one of which entered a cut in the mountains southward, while the other (of which was I,) continued on southeastward, and on the 17th crossed a mountain to a small stream tributary to the Jefferson. In the evening a Pen-d'oreille from the other division, joined us and reported that he had seen traces of a party of footmen, apparently following our trail. We ourselves saw during our march, the recent encampment of a band of horsemen, and other indications of the vicinity of probable foes. Pursuing our route, on the following day we reached and descended into the valley of the Jefferson twenty-five miles below the forks. This valley extended below us fifteen or twenty miles to the northward, where the river bending to the East, enters a narrow passage in the mountain between walls of cut rock. The plains are from two to five miles in breadth, and are covered with prickly pear, - immediately bordering the river are broad fertile bottoms, studded with cottonwood trees. The River is about one hundred yards wide, is clear, and has a gentle current,- its course is northward till it leaves the valley. We found the plains alive with buffalo, of which we killed great numbers, and our camp was consequently once more graced with piles of meat, which gave it something the appearance of a well stored market place. From starvation to such abundance the change was great, and the effect was speedily apparent. Indians, children, and dogs lay sprawling about, scarcely able to move, so gorged were they with the rich repast, the first full meal which they had, perhaps, enjoyed for weeks. The squaws alone were busy, and they having all the labour of domestic duty to perform, are seldom idle. Some were seen seated before their lodges with buffalo skins spread out before them, to receive the fat flakes of meat they sliced for drying. Others were engaged in procuring fuel, preparing scaffolds, and making other preparations for curing and preserving the fortunate supply of provisions thus obtained. Even the children were unusually quiet and peaceable, and all would have been exempt from care or uneasiness, had not the unslumbering cautiousness of the veteran braves discovered traces of lurking enemies.

On the morning of the 19th several of our men returned from their traps, bearing the dead body of Frasier, one of our best hunters, who went out the day previous to set his trap, and by his not returning at night, excited some alarm for his safety. His body was found in the Jefferson, about five miles below camp, near a trap, which it is supposed he was in the act of setting when fired upon. He was shot in the thigh and through the neck, and twice stabbed in the breast. His body was stripped, and left in the water, but unscalped. - In the afternoon we dug his grave with an axe and frying pan, the only implements we had that could be employed to advantage in this melancholy task, and prepared for the sad ceremony of committing to the earth the remains of a comrade, who but yestermorn was among us in high health, gay, cheerful, thoughtless, and dreaming of nothing but pleasure and content in the midst of relations and friends. Having no coffin, nor the means to make one, we covered his body in a piece of new scarlet cloth, around which a blanket and several buffalo robes were then wrapped and lashed firmly. The body thus enveloped was carefully laid in the open grave, and a wooden cross in token of his catholic faith placed upon his breast. Then there was a pause. The friends and comrades of the departed trapper gathered around to shed the silent tear of pity and affection over a companion so untimely cut off; and the breeze as if in sympathy with their sorrow, sighed through the leaves and branches of an aged cottonwood, which spread its hoary and umbrageous arms above his last resting place, as though to protect it from intrusion; while in contrast with this solemnity merry warblers skipped lightly from limb to limb, tuning their little pipes to lively strains, unmindful of the touching and impressive scene beneath. At length the simple rite was finished, the grave closed, and with saddened countenances and heavy hearts the little herd of mourners retired to their respective lodges, where more than one of our ordinarily daring and thoughtless hunters, thus admonished of the uncertainty of life, held serious self-communion, and perhaps resolved to make better preparations for an event that might come at almost any moment, after which there can be no repentance. But it may be doubted if these resolutions were long remembered. They soon recovered their light heartedness, and were as indifferent, reckless, and mercurial as ever. - Frasier was an Iroquois from St. Regis, in Upper Canada. He left that country seventeen years before, having with many others engaged in the service of the Norwest Company, and came to the Rocky Mountains. Subsequently he joined the American hunters, married a squaw by whom he had several children, purchased horses and traps, and finally as one of the Freemen led an independent and roving life. He could read and write in his own language, was upright and fair in all his dealings, and very generally esteemed and respected by his companions.

It commenced raining in the afternoon of the following day, and continued without intermission during the night. Taking advantage of the storm and darkness, a party of Blackfeet boldly entered our lines, and cut loose several horses from the very centre of the camp. An alarm having been given the Flathead chief arose and harrangued his followers, calling upon them to get up and prepare to oppose their enemies, not doubting but that an attack would be made at day break. When he had concluded, a Blackfoot chief, who last summer deserted from his people and joined the Flatheads, in a loud voice and in his native tongue, invited all who were lurking about camp, to come in and help themselves to whatever horses they had a mind to, asserting that as the whites and Flatheads were all asleep, there could be no hazard in the undertaking. Scarcely had he done speaking, when the Blackfeet, to testify their gratitude and appreciation of this disinterested advice fired a volley upon him. Fortunately, however, no one was injured by the firing, though several lodges were perforated by their balls. In the morning we were early on the alert, but the Blackfeet had all departed, taking with them seven or eight of our best horses. As there was no help for it, we had to put up with the loss, and the next day having finished drying meat, we struck our tents, and departed southward up the Jefferson.

Previous to our reaching this river we had exacted a promise from the Indians to accompany us to the three forks of the Missouri, but since the death of Frasier they refused to fulfill their engagement, asserting that we shall certainly fall in with a village of Blackfeet, who will dispute with us every inch of ground, and thus render the expedition to no purpose, for trappers would forget their employment when death was grinning at them from every tree and cluster of willows. Our route was therefore necessarily somewhat changed, and on the 23rd we reached the Philanthropy, and halted two or three miles from its mouth. This is a deep muddy stream thirty paces in breadth, flowing for the last twelve or fifteen miles of its course through an open valley, and finally discharging itself into the Jefferson, which it enters from the northeast, a short distance from Wisdom River, a branch proceeding from the Big Hole. All these streams are bordered by fine grass bottoms, and groves of trees and willows. Six miles above the forks, on the west side of the Jefferson, there is a bluff or point of a high plain jutting into the valley to the brink of the river, which bears some resemblance to a beaver's head, and goes by that name. Hence the plains of the Jefferson are sometimes called the Valley of Beaver Head. These plains are everywhere covered with prickly pear, which constitutes one of the greatest evils - Indians aside - that we have to encounter in this country where moccasins are universally worn. The thorns of the prickly pear are sharp as needles, and penetrate our feet through the best of mocassins; they are extremely painful and often difficult to extract. In the evening we were joined by a Nezperce Indian who brought intelligence that the Rocky Mountain Fur Company were encamped in the Big Hole.

Chapter XXI

On the 24th we moved up the Philanthropy a few miles, and killed numbers of buffalo, which were numerous in all directions. In the afternoon a party of strange mounted Indians came into the plain in pursuit of a herd of buffalo, but discovering our camp fled precipitately to the mountains. We were joined in the evening by twenty-five lodges of Nezperces. For several days nothing of interest occurred. On the 27th we followed the course of the river through a narrow defile of a mile in length, and descended into an open valley which we found covered with buffalo. The old chief immediately encamped and desired that no person should leave camp for that day, but remain and rest the horses, as by so doing they would be able to hunt the buffalo the next morning to much better advantage. His directions were complied with, as it was necessary to lay in a supply of meat for future use, and with fresh horses much greater execution could be done than if they were fatigued. The doomed bisons were therefore allowed a few hours respite.

An Indian about noon brought us a note from Jervais, a partner in the Rocky Mountain Fur Company, stating that he was left with three men to trade with the Nezperces; that his partners had gone northwest to hunt the sources of Clark's River, and that Fitzpatrick, one of the partners in that Company, had been killed on his way to St. Louis with one companion the spring previous. Fraeb, who was with Jervais the fall before, left Cache valley in August for St. Louis in order to bring out an equipment next spring. The object of the note was to inform our Freemen where little conveniences could be procured in exchange for furs. In the course of the afternoon a party of horsemen boldly entered the valley, but quickly perceiving the danger fled for their lives. The Flatheads were speedily mounted and in pursuit, but with the exception of one who was overtaken and killed, they gained the mountains in safety. During the night a fire was kindled on the neighbouring mountain, and we heard the reports of several guns in that direction, but the Indians did not approach our camp.

On the 28th we passed the body of an Indian killed the day before, and the squaws agreeable to an ancient custom, gave it repeated blows as they went by. It was totally naked, scalped, and pinned to the ground by an arrow through the heart. Beside it lay a half worn garment, in which we recognized the pantaloons worn by Richards when he was killed in the spring. It was hence conjectured that this Blackfoot had a hand in the murder. If so the bloody deed was in part avenged, for his bones were left to moulder here, as were those of poor Richards near Kamas prairie. A party of our trappers, to day, a few miles from camp, discovered an Indian on the summit of a mound who beckoned them to come to him, and disappeared behind the hill. They wisely declined a more intimate acquaintance, and returned to camp without further investigation. It was probably a decoy to an ambush.

After laying in a sufficient store of dried buffalo meat, we passed southward, over ranges of prairie hills to a small stream that flows into the Jefferson below the

Rattle Snake cliffs. There the Indians left us on the third of October, and we, continuing our journey, passed down the stream to its mouth, and thence up the Jefferson through the Rattle Snake Cliffs to the forks where Lewis and Clark left their canoes. One of these streams rises with the sources of the Madison and Kamas Creek, and flows northwestward to its junction with the other, which has its rise in Horse Prairie. Ascending the latter two miles above its mouth, we entered Horse Prairie at a narrow gap between two high points of plains. Here we found the Flatheads from whom we separated on the east fork of Salmon River, with the trader Jervais and several "engages," (hired men.)

On the 8th two of our men accompanied by three or four Indians departed for the Trois Tetons, to meet Mr. Dripps who was expected this fall from the Council Bluffs, with an equipment of men, horses, and merchandise. The same day two Indians came to us from the band which left us on the third. They stated that a large party of mounted Blackfeet came near them on the sixth, but departed without firing a gun, probably awed by their numbers. We left Horse Prairie on the eleventh, and crossed the mountains westward to the east fork of Salmon River, following the same trail that guided Lewis and Clark there so many years before Us. Here we fell in with another village of Nezperces, whom we had not before seen. Accompanied by these Indians we continued down Salmon River to the forks, about twenty miles, and thence six or eight miles to an abrupt bend westward where the river, leaving the valley enters a dark passage through rugged mountains, impassable for horsemen. The valley of the Salmon River is separated from the Big Hole, to which we crossed, by a mountain capped with a succession of bleak points of naked granite, the stern majesty of which makes an impression upon the beholder such as few scenes of earthly grandeur can equal. On the 29th the Rocky Mountain Fur Company returned, having finished their hunt on the waters of the Missouri without molestation from the Indians. Shortly after leaving Cache valley, however, they were attacked on the Blackfoot by a large party of the enemy. The attack was made at day break, immediately after the horses were turned loose, which was unusually early. It was still so dark that neither party could see the sights on their guns, and hence they overshot each other, doing little mischief on either side. As soon as the firing commenced, the horses broke into camp and were refastened to their pickets. The Indians, finding that they should get nothing by fighting, resolved to try what could be effected by begging. A party then marched coolly up to camp and announced themselves Creas. "They said," says my informant, "that they mistook us for Snakes and professed to be very sorry that they had commenced firing before ascertaining who we really were. Not a few of us raised our guns to punish their unparalleled impudence, but were restrained by our leaders, who believed or affected to believe their improbable story. We ascertained from them that the party was composed of one hundred Blackfeet and thirty-three Creas, and that several of them were slightly wounded in the fray. Our leaders made them a present, and suffered them to depart in peace much against the wishes of some of our exasperated men. Two of our trappers, who were absent from camp at the time of the attack never returned, and were doubtless killed by them. This occurred on the 15th of August."

On the 19th of the same month, four men (D. Carson, H. Phelps, Thos. Quigley and J. M. Hunter) left camp in Gray's Hole, and proceeded down Gray's Creek, in quest of beaver, about fifteen miles; during the time occupied in going this distance, they had set all their traps, and found the day too far spent, to look for a safe encampment, which is a rare thing here at best; however they halted near the brink of the river, where the margin was partially decked with here and there a lone cluster of willows, or birch, with some few intervening rose briars. The bottom or level margin of the river, extended but a few paces from the water's edge, and was there terminated by abrupt rocky hills, of considerable height, overlooking the bottoms, as well as the surrounding country, to a great distance. "We lay as much concealed as possible, in such an open place," says one of these men, whose account was corroborated by all the others; "and passed the night without disturbance; but just at day break, our ears were saluted with the shrill noise of the warrior's whistle, quickly answered by the re-echoing yells of a multitude of Indians, who were rushing upon us. We sprang from our beds, and in a twinkling one of our guns was discharged in their faces, which somewhat dampened their ardour, and they fell back a few paces; at the same time we sprang into the best position, the place afforded; the Indians re-appeared the next instant, and poured showers of lead and arrows around us. We saw no means of avoiding death, but resolved to sell our lives as dearly as possible. We mutually encouraged each other, and resolved if practicable, to fire but one gun at a time and wait until it was reloaded before firing again, unless the Indians should rush upon us, in which case we were to single out each one his man and send them before us to eternity. In short, each time they approached, the foremost was made to bite the dust, and the others fled precipitately; they were recalled, however, by the animated voice of a chieftain, who induced them to charge, time after time, upon us, but each time they advanced, the dying groans of a companion so completely unmanned them, that they fell back, again and again. At length, finding that they could not dislodge us, they fired upon and killed our restless horses, who were fastened a few paces from us, save one, which broke loose, and fell into their hands alive. In the mean time, others commenced throwing stones which fell thick around us, but fortunately did us no injury. After some time they departed, and ascended a high rocky hill some distance from us, where one of them stepped out before the rest, waved his robe five times in the air and dropped it to the ground, he then took it and disappeared with the others behind the hill. We immediately collected our blankets, saddles, &c. together with some articles the Indians had left, and concealed them as well as we could, intending to return for them, and set out for camp, which we reached without accident the same evening."

Early the next morning, a strong party set out with these men, to aid them in collecting their traps and baggage, but the Indians had already carried off every thing. They examined the battle ground and found several places where the ground was soaked with blood, and wads of buffalo wool were strewed about clogged with blood, with which they had stopped their wounds; trains of blood likewise marked the route of the fugitives, to twenty four stone pens where they had slept, which were mostly covered with proofs of the number of dead or wounded, that had lain in them. The persons, who visited the place, say that they cannot conceive how four men could be placed so as to escape death, where they were situated. The ground was literally ploughed up by balls, and all acknowledge that it was one of the most extraordinary escapes, ever heard of. The Indians were the same who attacked camp on the 15th. There were one hundred and thirty-three of them. The battle lasted from day break until ten o'clock, and these men fired about thirty shots, most of which were supposed to have taken effect.

Chapter XXII

A day or two after the arrival of the Rocky Mountain Fur Company, our men who were despatched about a month since to meet Dripps, returned, and reported that he had not reached the place appointed, but that Fraeb, who started for St. Louis last summer, fell in with Fitzpatrick, on the Platte, at the head of thirty men with pack horses. Fraeb immediately headed the expedition, which he was now conducting to this place, whilst Fitzpatrick returned to St. Louis, to bring out an equipment in the spring.

This last enterprising gentleman, departed in the month of February last, though necessarily exposed to every privation and hardship, to cross the whole extent of that immense plain, from the Rocky Mountains to the state of Missouri; which must needs be performed on foot, and a great part of the way on snow shoes, at that dreary season of the year. So hardy was this enterprise esteemed, that it was a matter of considerable speculation, among the brave Mountaineers, whether he would reach his place of destination, or not! He had promised, in case he should reach that place in safety; to bring an equipment to his partners in Cache Valley by or before the first of July. They awaited his arrival a month after the time had expired, and the opinion became universal, that he had been killed, or perished on the way. He reached the settlements, after a series of sufferings, and ascertained that his patrons Smith, Sublette, and Jackson, had left the state of Missouri two days before with a large assortment of goods for Santa Fe. Notwithstanding the fatigue our traveller had already undergone, he immediately procured a horse, and again entered the uninhabited prairies in pursuit of his friends, whom he overtook after several days hard riding. They persuaded him to go on with them to New Mexico, promising to give him an equipment at Toas, which would not be more than twenty days march from Cache Valley, whither he could arrive in time to meet his companions in the month of July.

Several days after his arrival among them, the party was charged upon by several hundred Comanche Indians; however, they were so terrified at the discharge of a six pounder, that they fled in alarm and adjourned the attack sine die. Shortly after this, one of the leaders of the party, Mr. Jerediah Smith, (a gentleman, whose life for several years in the Rocky Mountains, was a constant series of bold adventures, defeats, narrow escapes, and attendant miseries,) was killed during a lone excursion in search of water, for want of which, the party suffered two days, a thirst rising nearly to madness. A young man, employed by the company as clerk, whose name I did not learn, was likewise killed about the same time.

A few days after the last event, a large party of "Gross Ventry of the Prairie," encamped around them, but betrayed no evil intentions. The Chief said that he had buried all his resentment towards the whites, and should never annoy them any more. Probably the appearance of one hundred men, well armed, in a camp well fortified by the waggons and baggage added to the ever primed big gun continually pointed towards them, produced this salutary, though perhaps temporary effect.

The party reached Toas, on the Rio del Norte; and Fitzpatrick having received his equipment, departed for the mountains; but being unacquainted with the route, and having no guide, he missed his way, and fell on to the Platte, where he met with Fraeb as before mentioned. Fraeb met also on that river with a party of fifty men, led by a Capt. Ghant. They were all on foot, and led about their own number of pack horses, and were destined for the mountains.

Two days after our express returned, three others of our men who were confident that Dripps would come on this fall, set off to meet him. Fraeb arrived one or two days after their departure, and camp presented a confused scene of rioting, and debauchery for several days, after which however, the kegs of alcohol were again bunged, and all became tranquil.

The men provided themselves with lodges, and made preparation for passing the winter as comfortable as possible. We purchased all the dried meat the Indians could spare, together with robes, and "appishimous" (square pieces of robes, used under our saddles in travelling, or under our beds in camp,) in addition to our former stock of bedding. Our arrangements completed, we had nothing to do, but to make the time pass as easily as possible. We assembled at each others lodges, and spent the evening merrily, by listening to good humoured stories, and feasting on the best the country afforded, with the frequent addition of a large kettle of coffee, and cakes.

On the 6th of November, one of the three men who departed sometime since, to meet Dripps; returned, and reported that himself and comrades had been east of Snake River, but, that during their journey, they had seen several war parties of foot Indians who pursued them until they finally resolved to return, fearing that they would discover their encampment some night, and steal their horses, if not their lives. On the evening of the third day of their journey homeward they encamped in a dense thicket of willows, on the east fork of Salmon River, where they imagined themselves quite secure; but the following morning, a rustling of leaves and brushes, betrayed the approach of something unusual. They immediately sprang from their beds, and by this movement, discovered their place of concealment to the wary Indians, who now commenced firing upon them. One of them Baptiste Menard, was soon severely wounded in his thigh, and his groans served to increase the ardour of the enemy, who now pressed forward with resolution; but the first who presented himself was sent to the other world, by a well directed shot, which at once put an end to the action. The Indians lost all their courage with their friend, and immediately departed, taking the horses with them. After they were gone, our men conveyed their wounded comrade a mile or two, to a place of more security, and remained until dark, when my informant departed to get assistance from camp. He had not proceeded far, however, when the Indians discovered him, and gave chase, but he escaped in a thicket of willows; and thence continued his progress, without interruption, until he reached camp, which he did the next evening; having walked fifty miles since he left his companions. The morning after his return, a party of volunteers set out for the wounded man and his companions and returned with them on the third day afterwards. This man Menard, was shot in the hip and the bones so fractured that he remained a cripple for life.

About this time, a large party of Flatheads, and others, departed for buffalo, promising to return in the coming moon. Two or three days after, one of them returned with the news, that they had recovered some stolen horses from a party of Blackfeet, and taken two of their scalps. On the 21st of December, two men from Mr. Work's party, (Hudson Bay Company) arrived and stated that Mr. Work was encamped two day's journey above, on the east fork. They had been to Beaver Head, and were continually harrassed by the Blackfeet, who killed two of them, and severely wounded a third. They killed, however, several of the enemy, and captured a number of horses. They saw the body of a man in the Jefferson River, below Beaver Head, which our hunters believed to be the body of Frazier, whom we had buried there.

On the 23rd we separated from the Rocky Mountain Fur Company, and passed southward up Salmon River, to the western extremity of little Salmon River valley, forty miles above the entrance of the east fork.

The river was all the way confined by lofty mountains on either side, and numerous points jutting into it, rendered the journey extremely toilsome, for our jaded horses. However, our difficulties ended when we entered the valley, though we continued twenty miles up it, and encamped with a few lodges of Flatheads, on the 3d of January 1832. In this valley we killed upwards of an hundred head of buffalo, which were numerous for sometime after we arrived. Heretofore the weather has been warm, and pleasant during the day time, but the nights extremely cold. The rivers have been frozen for a month past, but the valleys are still free from snow.

I departed with three others on the 25th, to procure some trifling articles from the Rocky Mountain Fur Company. We returned down Salmon River, and reached a village of Nezperce Indians, late in the evening of the second day, with whom we remained one night. The hospitable Indian I chanced to stay with, treated me with great kindness, and contrary to my expectation refused any remuneration whatever. From him I learned that the Rocky Mountain Fur Company were encamped twenty miles above, on the east fork, together with forty or fifty lodges of Flatheads, and Nezperce's. We continued our journey the next morning, and reached them late in the afternoon. They were encamped opposite to a pass to Horse Prairie, well known to the Blackfeet, who had lately stolen twenty horses, and fled by that route to the Missouri.

The second evening after we arrived, soon after dark, a party of Blackfeet approached camp, and several of them boldly entered, at different points, cutting loose our horses in their way. One of them mounted a beautiful horse, and slowly rode through both encampments. During his progress he was challenged by the guard, but gave the usual Flat Head answer and passed on; soon after his departure, the owner of the horse discovered that he was missing, and imagining that he had broken loose, departed with a companion in quest of him. They proceeded silently about fifty yards from camp, and met a Blackfoot who came running up to them, thinking they were some of his comrades; but quickly discovered his mistake and fled. They brought him to the ground, however, by a well-directed shot, and about twenty others immediately sprang up from the sage, and fled into the woods bordering the river. The Flat Heads raised the scalp of the dead Indian, by cutting around the edge of the hair and pulling off the entire skin of the head from the ears up. The taking, or raising of a scalp is done in this way, by all the mountain tribes. We ascertained next morning that the Blackfeet had taken seven or eight horses. The Indian killed, as stated above, was a tall, bold-featured, handsome fellow, unusually white, and about twenty-two years of age.

Two days after this affair, an express arrived from Mr. Work's party, who were at this time with a large band of Pen-d'oreilles, at Beaver Head; they had lost several horses, which were stolen by the Blackfeet, and had hemmed up a body of those Indians, so that neither party could injure the other; but could yet talk freely on both sides. The Blackfeet stated that the white chief, at the mouth of the Yellow Stone River, (McKensie of the A.M.F. Co.) had built a trading house at the mouth of the Maria; and had already supplied the Blackfeet, with one hundred and sixty guns and plenty of ammunition; and they were now, only awaiting the arrival of a large band of Blood Indians from the north, to commence a general war of extermination of all the whites, Flat Heads and others in this part of the country. The day after the express arrived, I departed with my companions, and reached our own quarters without accident about the third or fourth of February.

Chapter XXIII

The fine valley in which our camp was situated, is thirty miles long, and twelve broad; it is intersected by willowed streams, and large bottoms, covered with rich pasturage, hence it is a favourite resort for both deer and buffalo. The only trees are a few orchard-like groves in the head of the valley, and pines of every variety, on the abrupt sides of the surrounding mountains. The principal stream flows northwestward into Salmon River, which runs northward through the lower extremity of the valley. On the 9th of February, we passed up to the head of the valley and left the Indians, who had hitherto accompanied us behind. Previous to this time, we had scarcely seen a particle of snow in this valley; but we were now detained, by a snow storm of four days continuance, which left the lowlands covered to the depth of one foot. However on the 15th we passed through Day's defile, where we found the snow two feet deep, and covered with an icy crust that cut our horses' legs so that they bled profusely. We proceeded slowly, and employed our best horses successively to break the road, until we reached a small patch of willow on Day's Creek in view of the plains of Snake River. The day was intensely cold, and many of us frost bitten, notwithstanding we had taken the precaution to envelope ourselves with blankets, and buffalo robes. At this evening's encampment, we found nothing but small willows for fuel, and even a scarcity of them. At midnight we were brought to our feet by the cry of Indians, and sprang out to our horses, twenty-five of which were missing. We saw several pairs of snow shoes, and as many packed dogs, but the Indians had vanished with our horses, and the night was so extremely cold, that no person could be induced to follow them, though we had every reason to believe that they could be soon overtaken. Many of our companions intended to set out in pursuit at day break, but the drifting snow so completely erased all trace of the robbers, that no one could designate the course they had taken.

After this period, we continued slowly down the extremity of Day's Creek, whence we were in full view of the Trois Tetons; the three buttes of Snake River, and the mountains east of the river. The three buttes of Snake River, are three gigantic, solitary, or isolated mounds, rising from the plains, midway from Snake River, near the mouth of Porteneuf, to the mountains northward. They are fifteen or twenty miles asunder, and the most westerly richly deserve the name of mountain. It is covered with pines, abounds with big horns, and is crowned with snow almost the year around.

From the extremity of Day's Creek, we continued southward in the direction of the middle butte, fifteen miles to Gordiez River, which was quite dry when we reached it. There were several cotton wood trees scattered along the margin, but none of those long grass bottoms, common to other streams are to be found here: in lieu of them a sandy uneven plain appears, covered with black rocks, and wormwood, extending as far as the eye can reach, and likewise covered with pools of water from the melting snow, which is rapidly disappearing. We saw near the margin of the river, the trail of an Indian village that had passed two or three days since to the westward. Fifteen of our party immediately set out in pursuit of them, hoping to hear something of our stolen horses. The following day we passed under the south side of the middle butte, and encamped in a large grove of cedars, two miles from Gordiez River. The next day we continued about the same distance, and halted in the sage on the open plains. We saw large herds of buffalo during our march, and killed several, which to our surprise, were as fat as they generally are in the summer season. In the evening, two hundred Indians passed our camp, on their way to the village, which was situated on the lower butte. They were Ponacks, as they are called by the hunters, or Po-nah-ke as they call themselves. They were generally mounted on poor jaded horses, and were illy clad with shirts and leggins, of dirty torn or patched skins, moccasins made of buffalo skins, and old buffalo robes, half divested of hair, loosely thrown over the shoulders, and fastened by a string around the middle. They were generally ugly, and made a wretched appearance, illy comparing with their bold, handsome and well clad neighbours, the Flatheads. They gave us to understand, that a party of whites were now in Cache valley. On further enquiry, we were satisfied that it could be none other than Dripps, who we supposed had got thus far, on his way to Salmon River last fall, but was prevented from continuing his journey, by the bad condition of his horses, and almost total want of grass on the route.

The next day we reached Snake River, opposite to the mouth of Blackfoot. The same evening the party of fifteen who left us on Gordiez River, returned, having gained no information of their horses. They went to the village of Ponacks at the western butte, and represent them to be miserable, in the superlative sense of the word.

On the 4th of March we crossed the river on the ice, and encamped near the mouth of Blackfoot. The plains are now entirely free from snow, though they are not dry. On the 5th John Gray and David Montgomery, departed for cache valley, to ascertain if Dripps was there, or not. A day or two afterwards, the Ponacks came and encamped a short distance below us. On the 10th we left our thriving neighbourhood, and halted at a spring east of Porteneuf: - the same evening two of our hunters brought in Gray, (one of two men who left us on the 5th) whom they found lying half dead in the cedars, near Porteneuf. He gave us the following account of his unfortunate trip to Cache Valley.

"We proceeded," said he, "by way of the south fork of Porteneuf to Cache Valley, without accident, and sought throughout the northern extremity, for traces of the whites, but were unable to find the least evidence of their having been there at any time during the winter. Hence we concluded, that the story told to us by the Ponacks, was a falsehood invented solely to draw from us a present, which is usually given to Indians on the receipt of good news. This conviction added to numberless traces of foot Indians, that appeared wherever we went, induced us to return back to camp with the least possible delay. In the afternoon of the 8th we discovered a small herd of buffalo, and succeeded in killing one of them, after firing several ineffectual shots. Our appetites had been quickened by two days starvation, which urged the adoption of bold and prompt measures. We quickly secured the tongue, with other choice pieces, and proceeded in quest of fuel, at a rapid pace. During our progress, we saw what greatly resembled an Indian, laying upon the ground, with his buffalo robe thrown over him. We hesitated a moment, but concluded it to be the carcase of a buffalo, and continued on. At length, we reached a small lake, which is the source of the south fork of Porteneuf. It had been frozen over in the early part of the winter, and was since covered with water to the depth of one foot, which was encrusted with a sheet of ice, though not strong enough to bear one. Near the margin, were several clusters of large willows, which were now surrounded by ice and water; they supplied us with fuel, which we conveyed to the bank, beyond the reach of the water, and kindled a fire, by which we roasted and devoured our meat, with tiger-like voracity, until our hunger was allayed.

"By this time the sun was disappearing behind the western hills, and being fully aware of the danger of remaining in such an open place all night, I remarked to Montgomery that we had better saddle our horses, and proceed down the creek, until after dark, and pass the night in some of the groves of cedars which were scattered along the entrance of ravines, in our route. He objected to this measure, and added that wiser men than ourselves had encamped in worse places. Finding that remonstrance would be useless, I immediately cut away some of the briars in the centre of a bed of wild rose bushes, and spread down our blankets. At dark we lay down, and my companion slept soundly. For my own part, I was alarmed in the early part of the night by some unusual noise, which might have been occasioned by the trampling of our horses; but which, together with a train of thoughts foreboding evil, effectually prevented me from closing my eyes to sleep at all.

"I arose early in the morning, but it was yet light, and commenced kindling a fire, in the course of which, having occasion for my powder horn, I called to Montgomery to hand it to me. He immediately arose and stepped out, but sprang back to his bed the next instant exclaiming Indians! Indians! At one bound I was with him, and the Indians commenced firing upon us. The rose bushes which surrounded us, only served to conceal us from view but offered no resistence to their balls, one of which grazed my neck. I immediately exclaimed "Montgomery I am wounded." The next instant he arose with his gun to his face, in a sitting posture, but ere he had time to shoot, his gun dropped from his hands, streams of blood gushed from his mouth and nose, he fell backwards uttering a groan, and expired. I sprang up, and presented my gun to the advancing Indians, determined to kill one of them, but they threw themselves down in the grass. I then wheeled and fled through the breaking ice of the lake, and exerted my utmost strength, to gain the opposite bank. Some of the Indians were instantly in close pursuit, whilst others deliberately fired from the bank. One of their balls grazed my thigh and another cut out a lock of my hair, and stunned me so much that I could with difficulty keep my feet; however, I succeeded in reaching the bank, but had the mortification to see the foremost of my pursuers step ashore as soon as I did. At this moment, a thought crossed my mind, to surrender all I had and they would spare my life; but the recollection of the cruelties they have ever practiced upon prisoners, always terminating in death, awoke me to reason, and I redoubled my efforts to gain a ravine, which led into the mountain. As I reached the entrance, the loud, harsh voice of the chief, calling back my pursuers, fell upon my ears like strains of the sweetest music; but I continued running until overcome by exertion, I fell down quite exhausted. After resting a few moments, I ascended the mountains and dragged myself through the snow until dark, in the direction of Snake River, at which time, I descended to the margin of Porteneuf, and followed its course.

"My mocasins became worn out and left my naked feet to be cut and lacerated by the ice and stones, and at the same time, I was drenched by a shower, which chilled me through. I endeavored to kindle a fire, and make use of the powder in my gun for the purpose, but was unsuccessful. There being no alternative, I was compelled to crawl along or freeze. My feet, now became extremely painful, and I found they were frozen. Being no longer able to support myself upon them, I sought a stick with which I hobbled along some distance, but at length found myself in a field of prickly pears, that pierced me to the very soul. Here, for the first time, I wished for death and upbraided myself for running from the Indians. I stopped and plucked the thorns from my bloody feet, proceeded and the next moment was again upon them. At length, I crawled into the willows, bordering the river, and to my great joy found a quantity of bull rushes. Fortunately, I happened to have a pen knife, with which I cut as many as I could grasp in my arms twice, and bound into three separate bundles; these I fastened together with willows, launched it without difficulty, and embarked upon it, allowing it to be carried along by the force of the current.

"In the afternoon of the following day, I reached the nearest point from Porteneuf to camp, and abandoned my floating bed. With a stick in one hand and my gun in the other, I set out; but the torture from my feet was such, that I fell down, unable to proceed farther. In this situation, whilst revolving in my own mind the chances for getting to camp, a distance of twelve miles, I was discovered by the two hunters whose presence gave me a thrilling sensation of joyful deliverence, indescribable. One of them immediately dismounted, and placed me upon his horse, which he slowly led to camp."

When Gray reached his own lodge, his mangled frozen feet were examined; they were swollen to twice their natural size, and were quite black; however, at the expiration of two months, he was quite well, and the circumstances of his so narrow escape almost forgotten. He left his powder horn, shot-pouch, belt, and knife at the field of death, which will account for his want of success, when endeavoring to kindle a fire; and for being compelled to construct his raft with a pen knife, which is a rare instrument in this country, because it is useless, save in such a peculiar case.

Chapter XXIV

After Gray's return, we moved camp over to Porteneuf. This stream rises between Blackfoot, and the Sheep Rock of Bear River, and flows fifty or sixty miles westward, to its junction with Snake River. On the south side, a point of mountains juts down nearly to Snake River; but on the north side, the mountains disappear. Fifteen miles above its mouth, the river enters the plains, through a narrow opening in the mountains, somewhat resembling a huge gate way, hence it is called Porteneuf, (New gate.) The banks of this stream are garnished with impenetrable thickets of willow, briars, and vines, matted together; bluff ledges of rock, where the country has evidently sunk, and here and there near the fork, remains of boiling springs. After this period, we continued to the source of the south fork of Porteneuf, and on the evening of the eighteenth, reached the spot where Montgomery was killed; the blood appeared quite fresh on the grass, where he had lain, but nothing could be found of his remains, save a few small bones. In justice to the memory of a careless, good-natured, brave, but unfortunate comrade, we resolved to call the pass, from Cache Valley to Porteneuf, "Montgomery's Pass."

On the twentieth we reached Bear River in Cache Valley, having seen during our journey, traces of foot Indians. Some of our hunters saw twenty Indians some distance from camp in the valley. On the twenty-third, several hunters arrived from a company of fifty, who had passed the winter in the southern extremity of this valley, and were now encamped a few miles east of us. This party was fitted out at Fort Union, at the mouth of the Yellow Stone; and was led by a Mr. Vanderburgh. Four of their men were killed in Cache Valley, during the winter, and as many others left them in the fall, but never returned. They were well-supplied with meat during the winter, and never had occasion to go down to the lower end of the valley; hence the reason why Gray and Montgomery did not fall in with them. From them we ascertained that a certain district where we intended to make our hunt, had already been trapped by a party from Toas, last fall. This information induced us to join Vanderburgh, and proceed with him forty miles northward, up Bear River, to the Sheep rock. This river was confined all the way, by cedar-covered or prairie-hills, and ledges of black rock.

The "Sheep Rock," is the high, rocky, abrupt termination of a mountain, south of the river, which flows around it, through a deep canal of cut rock, from the southeast. At the Sheep Rock is a beautiful cove or horse-shoe-like valley, two or three miles in diameter, bounded on the north and west by irregular hills, covered with fragments of black rock, and scattering cedars. From south to northeast, it is surrounded by lofty mountains, through which the river meanders, before it reaches the valley. There are groves of cedars in and about the cove, which likewise betrays an unusual volcanic appearance. The plain is covered, in many places, with a substance resembling ashes; the rocks have a black, blistered appearance, as if burnt; and there are the remains of many boiling springs similar to those on Salt River, which have long since exploded. Some of them present little knolls of a beautiful yellow, tasteless substance, several paces in extent; others present the hollow mounds of cement, that were formed by deposits from the waters, which have long since disappeared. There is a spring in the middle of the valley, the waters of which taste precisely like soda water, if drank after the effervescence has ceased. Some of these boiling springs were situated on the highest mounds, and others in the valley. We saw the skeletons of five persons bleaching in the grove of cedars, near the valley, supposed to be Indians. The country here is yet covered with water from the snows, which have just disappeared.

From the Sheep Rock, we followed the zig-zag course of the river seventy-five miles, and again entered a beautiful valley, fifteen miles long from north to south, and five or six broad; at the southern extremity, the outlet for the Little Lake enters, and falls into Bear River. The margins of both rivers are here decorated with dense groves of cottonwood and aspen trees, and thick underbrush, and the valley is a great resort for both animals and wildfowl, particularly geese, who always deposit their eggs in the old nests made by hawks and ravens, in the trees; great numbers of eggs are collected by passing trappers, in the spring. We reached this valley on the tenth of April; at this time our trappers branched out in various directions in quest of beaver.

On the thirteenth we continued twelve miles eastward, over prairie hills to Talma's Fork, a small stream that interlocks with the sources of Salt River, and flows southward into Bear River. It receives its name from an Iroquois who discovered it. Bear River has again meandered into a valley, at the mouth of Talma's Fork; thus far it varies from fifty to one hundred yards wide, is rapid and seldom fordable; its naked borders present nothing but an occasional lone cluster of willows, save in Cache Valley, and at the outlet of the Little Lake, where groves of trees beautify its margin.

On the fourteenth we passed eight miles southeastward, to Smith's Fork; this is a large well wooded creek, that rises with the sources of Ham's Fork and Salt River, and flows southeastward into Bear River. It commands a narrow valley until near its junction with the latter, where two high points of mountain, jutting towards it on either side, leave a narrow passage for the water. This stream is noted for the great numbers of beaver taken from it, and receives its name from the late Jerediah Smith, of the firm of Smith, Sublette and Jackson.

On the fifteenth we forded Bear River, at a place unusually shallow, passed twelve miles southeastward and re-encamped on its margin. From the south of Smith's Fork, the mountains, which have hemmed up the river more or less, since our departure from Cache Valley, expand, leaving an open plain five or six miles wide, bounded on the east by a high mountain, and on the west by a low one, which is abrupt on the western side, and overhangs the Little Lake. Through this plain, the river forms a gentle curve from east to south; the valley on the east side is apparently as level as the surface of still water; but on the western side, has a very gentle ascent, until it reaches the abrupt base of the mountain. The river is from fifty to eighty yards wide; is deep, and has a gentle current; its borders are in many places naked of bushes; but generally here and there, a solitary cluster of willows afford a resting place for the ravens, or a shelter for the wolves. The plains were graced with hundreds of antelopes, either gamboling about, or quietly feeding in groups, with ever watchful sentinels to apprise them of danger.

On the sixteenth we passed a few miles above the mouth of Muddy, and killed several buffalo from a large herd, which were the first we have seen since we left the valley, at the outlet of the Little Lake. We likewise saw great numbers of geese and ducks, which have just made their appearance in the river. On the twenty fourth we recrossed Bear River, and encamped on its eastern margin; during the afternoon a well known Flathead Indian, named Paseal, who accompanied Fontenelle and Dripps to St. Louis last summer, returned with the agreeable intelligence that Dripps, at the head of forty-eight men, was encamped at the entrance of Muddy. We moved down on the following day, and encamped with him: we now ascertained that he left the council Bluffs about the first of October, but owing to want of grass, and the jaded state of his horses, was compelled to stop, and pass the winter at the foot of the Black Hills. In the mean time he despatched three men and an Indian to us on Salmon River, who ought to have reached that place previous to our departure, but they have not been heard of since. Two or three of the following days were devoted, by many of the men to inebriation; a chilling storm of sleet, attended their out of door revels.

On the twenty-ninth I set out with three others, to raise a small cache of furs we had made on Rush Creek in Cache Valley. We proceeded by way of the Little Lake forty-five miles to the head of Cache Valley, and thence thirty-five, by night, to Rush Creek. This is a small stream (that flows into Bear River, on the south side,) bordered by dense thickets, and at this time was not fordable. I followed the brink several hundred yards, in hopes of finding a shoal, where we could cross without wetting our fur; at the same time one of my comrades who was mounted, entered the brush a short distance above me, for the same object. Soon after, hearing a noise like that of some large animal splashing in the water, I ran to the spot, certain that my comrade had attempted to cross, where the river was deep and his horse endangered. Imagine then my agony and surprise when a formidable grizzly bear came rushing, like a wounded buffalo towards me. I instinctively cocked my gun, and intended to discharge it into his open mouth, when he should rear himself to clasp me; but to my great joy he passed a few feet from me, and disappeared in the neighbouring thickets.

We returned the following night to the head of Cache Valley, and were saluted by the barking of several dogs during our route; however the night was dark, and we rode briskly until we were beyond the reach of either dogs or Indians. We suffered from exposure to a snow storm, of two or three days continuance, but at length reached camp at the mouth of Smith's Fork, after a march of five days and two nights.

Chapter XXV

On the eighth of May, we continued northwestward, down Bear River, and reached the Horse-shoe Cove on the twelfth. A mile or two above the Sheep Rock, and a few yards from the river, is a bed of chalk white substance, called "the white clay," which possesses the cleansing property of soap, and is used by the hunters as well as the natives, instead of that commodity. It is found in various parts of the country, and is sometimes called 'white earth.' On the following day we passed northeastward, through cedar hills, which opened into a plain, decked with groves of cedar, and bluff ledges of rock, where the country, or at least portions of it, have evidently sunk. In the course of our route, we frequently marched several miles over a level plain, and suddenly came to an abrupt precipice, twenty or thirty feet high, where we sought vainly to find a place sufficiently oblique to admit of descending without danger. When safe below, we continued our progress in like manner, over a level country some distance, until another precipice obstructed our progress. High lone mounds, rising out of level bottoms, are not uncommon. We encamped fifteen miles northeast of the Sheep Rock, on one of the sources of Blackfoot.

Near our encampment were found an American riding saddle, and a rifle that was stripped of the lock and mountings. These articles were recognized to have been the property of Alexander, one of four men, who left Vanderburgh near the Big Lake last fall. Heretofore it had been believed that they were killed by some of the Blackfeet, who were lurking about Cache Valley last fall and winter. That opinion was mournfully confirmed by the circumstance of finding these articles, eighty miles indeed from that place, but directly in the route of the Blackfeet to their own country. We likewise saw ten Indian forts in a grove of cedars, that had been but recently evacuated.

On the fourteenth we continued in the same direction, about the same distance, and halted at the brink of another source of Blackfoot. Previous to this time for several days, we have had raw disagreeable weather, but it is now quite pleasant. Buffalo and antelopes, have been continually in sight since we left Smith's Fork. Next day we passed northwestward, through a plain intersected by numbers of small streams, flowing through deep canals of cut rock, which unite and form Gray's Creek, which is likewise confined between barriers of cut rock. This valley, or rather district, is called Gray's Hole, after John Gray, a half breed Iroquois, who discovered it some years since. This person is the same who was with Montgomery when he was killed.

In a narrow bottom beneath the walls of Gray's Creek, we found a party of trappers, headed by Bridger, one of the partners in the R. M. F. Company. Their encampment was decked with hundreds of beaver skins, now drying in the sun. These valuable skins are always stretched in willow hoops, varying from eighteen inches, to three feet in diameter, according to the size of the skins, and have a reddish appearance on the flesh side, which is exposed to the sun. Our camps are always dotted with these red circles, in the trapping season, when the weather is fair. There were several hundred skins folded and tied up in packs, laying about their encampment, which bore good evidence to the industry of the trappers. They found a rifle, as well as ourselves, which was likewise robbed of the lock and mountings. It belonged to one of two men, who disappeared a day or two previous to the battle, in August last. Both of these rifles were unusually heavy, and were doubtless left by the Indians for that reason.

On the nineteenth I departed from camp, accompanied by two Indians, to seek the Flatheads, and induce them to come to the forks of Snake River, where our leaders wished to meet them, for the purpose of trading. We passed ten miles over rocky hills, to the plains of Snake River; thence fifteen, to the mouth of Gray's Creek, and forced our horses to swim over Snake River, which we crossed on a raft ourselves. We halted a short time on the western margin, to bait our horses, and again proceeded northwestward. Six miles from the river, we passed a small lake, which is the termination of Cammas Creek, and has no outlet. We continued our course four miles beyond the lake, and halted in the sage after dark without water. We started at daybreak on the twentieth, and directed our course towards Cotas defile. During our march, we saw great numbers of buffalo running in various directions, which convinced us that they had been alarmed by Indians. This startled us in no small degree for we did not doubt but that they were Blackfeet, and should they discover us in the open plains, escape with our jaded horses would be impracticable. However, after suffering a fever, occasioned by thirst and excitement, and marching thirty-five miles over the heated plains, we reached Cotas Creek, and gladly threw ourselves down to sip the refreshing waters that flow from fields of snow in view. Our minds, however, were not yet free from apprehension, for just before we reached the river, three horsemen appeared coming towards us at full speed; two of whom came near enough to satisfy themselves that we were certainly men, and then turned and fled up the river. We immediately cooked and eat several choice pieces of a buffalo we were fortunate enough to kill in the morning, and remained until dark watching by turns the appearance of Indians, but saw nothing save here and there a veteran bull, quietly feeding around us; or large herds of buffalo in the distance. At dark we saddled our horses, and departed cautiously up the river, carefully avoiding to ride near the margin. Soon after our departure, our horses turned towards the river, and neighed, a certain sign that they saw or smelled horses; we continued, however, without annoyance, about ten miles, and halted to pass the night on the steep side of a hill.

The next morning at daybreak, we were on the march, and passed through a narrow space between two bluff ledges of rock, into a large plain, where Cotas Creek, and the east fork of Salmon River, both take their rise. We continued twenty miles down the plain, when we discovered a large party of horsemen meeting us at full speed. We hastily ascended an eminence, unsheathed our guns, and with no little anxiety awaited their approach. As they came near, we hailed them in Flathead, and they immediately discharged their guns in the air, which relieved our minds at once from apprehension. We followed their example, and descended to them. They were Flatheads, and Nezperces, and had just started for buffalo; but after hearing our mission, they furnished us with fresh horses, and returned with us at half speed, about six miles, to the village. Here we found the men Dripps sent in quest of our party from the Black Hills last winter. They reached the village last spring, a few days after we left Salmon River.

The Indians had had a battle with the Blackfeet three days before I arrived. They lost twelve men killed, and several others severely, if not mortally, wounded; besides upwards of a thousand head of horses, which were taken by the Blackfeet. The latter left sixteen of their comrades dead on the field. The action lasted two days, and was so obstinate at the commencement, that six or eight of the Flathead tents were cut up by their enemies, and several of the latter killed in camp. There were about a thousand of the enemy, who came for the purpose of annihilating the Flatheads, root and branch. Previous to the commencement of the fray, they told the Flatheads that McKensie had supplied them with guns by the hundred, and ammunition proportionate, and they now came with the intention of fighting, until "they should get their stomachs full." After the battle, when as usual in such cases they were crying for the loss of their friends; the Flatheads demanded sarcastically, if they had "got their stomachs full," to which they made no reply, but immediately departed for their own country. Sixteen of their scalps were triumphantly displayed by the Flatheads, who courageously defended their own slain, and prevented the Blackfeet from taking a single scalp. Several of the Flathead horsemen were killed in the spring, previous to the battle, amongst whom was the brother of Pascal, one of the Indians who accompanied me. On the twenty-second we departed, and bore southeastward up the plain. The wounded Indians were carried on a kind of litter simply constructed, by fastening the ends of two long poles to opposite sides of a pack horse, and tying cross bars six feet assunder, to prevent the long poles from approaching to, or receding from each other. A buffalo robe is then fastened loosely to the four poles, and the wounded person placed upon it. These litters, of which there were eight or ten, were followed by numbers of young men, ever ready to administer to the wants of the sufferers. Among the latter, was a young man who was shot through the knee; - his leg was swelled to an enormous size, yet he would not allow himself to betray the least symptoms of pain, and exultingly gloried in his misfortune.

We reached the narrows at the head of the plain, and the source of Cotas Creek on the twenty-third. Considerable anxiety was now manifested by the Indians. They were without either provisions or ammunition, and were consequently only prevented from pushing forward, to where both could be obtained, by the inability of their wounded companions, to endure the torture occasioned by long marches.

On the twenty-fourth we passed down Cotas defile, and fell in with a party of Flatheads, who left the village previous to the battle. They were well supplied with both dry and fresh meat, and at the same time were surrounded by buffalo, numbers of which were killed by our party. These Indians were probably the same discovered by us, and believed to be Blackfeet, on our way up four days since. After this period we moved slowly down Cotas Creek as far as the mountains jut down into the plain, on either side, and killed numbers of buffalo, which were numerous in all directions. In the meantime three of the wounded Indians died, and were decently buried. They were enveloped in skins lashed around them, previous to interment, and their graves after being filled with earth, were surmounted by little comical heaps of stones, which is the only mark by which the resting place of these heroes may hereafter be designated.

Chapter XXVI

On the 2d of June, a party of hunters arrived from our own camp, which was situated a few miles above the forks of the Snake river. The following morning I departed in company with one of the hunters, for camp; we passed twenty miles North of East, through a sandy plain decked with great numbers of Rocky mounds, which were all cross cracked, at the top, leaving cavities in some cases, large enough to shelter both men and horses, from the balls or arrows of Indians. The largest are one hundred feet high, and overlook the country far, in every direction. They appear a secure asylum to small parties of men, who, if once within them, may bid defiance to hundreds of Indians. A mountain of white sand, thirty miles in extent, is situated six or eight miles north of the forks of the Snake River. I have crossed several points of it, with difficulty, owing to the depth my horse sank into the sand. In most places it is entirely destitute of all herbage, and at a distance resembles a snow clad mountain. We reached camp in the afternoon, and ascertained that nothing worthy of recollection, had occurred since I left it. The trappers were all in camp, having ceased to trap, and the Springs hunt was considered over.

The next day the Indians reached us, and were requested to accompany us to Pierre's Hole, where we expected to meet Fontenelle, with supplies from St. Louis. They agreed to accompany us, if we would remain with them a day or two, to rest their jaded horses. In the meantime the brave Indian who was shot through the knee, died, and was buried on the margin of Henrie's fork.

After this period we continued slowly up Henrie's fork, and halted two or three days on the East fork, to dry meat, knowing that we should remain one or two days at rendezvous, and that buffalo would soon be driven far from us. We killed hundreds daily during our stay on Henrie's fork; and continued thirty miles South Eastward over prairie hills, decked with groves of Aspen trees, to the Northern extremity of Pierre's Hole. This pleasant retreat is twenty miles long, and two wide, extending from South-east to North-west; and is surrounded by lofty mountains, save on the west side, where prairie hills appear. It is watered by numbers of small streams, which unite and form Pierre's fork, a fine stream thirty or forty paces in width, which cuts its way out of the valley, in a deep canal of bluff rocks. On the east side of the valley, three majestic peaks of naked rock, rise far above the rest, and are well known to mountain rovers by the name of "The Trois Tetons." The mountains are very abrupt, as far as the pines extend, and the huge pyramids above are absolutely inaccessible. This valley is noted for the large extent of excellent pasturage, along the borders of its waters; and has been selected as a pleasant place for a general rendezvous, by the R. M. F. C., Vanderburgh and ourselves: it receives its name from an Iroquois chieftain, who first discovered it; and was killed in 1827, on the source of the Jefferson River. On reaching this valley, we found the Rocky Mountain Fur Co. already here, awaiting the arrival of Mr. Fitzpatrick, with supplies from Saint Louis. Mr. Vanderburgh expected a Mr. Provenu, with an equipment from fort Union, at the mouth of the Yellow Stone; and we as anxiously looked forward for Mr. Fontenelle, who was expected from the Council Bluffs.

Some days after we entered Pierre's Hole, a party of trappers returned, having made their hunt to the Southward. They saw Captain Ghant, at the head of fifty or sixty men, on Green river; he had procured horses from the Spaniards of New

Mexico, and had made his hunt on the sources of the Arkansas, and tributaries of Green river, without molestation by the Indians. Two men were despatched by the R. M. F. Co. about this time, to meet the Saint Louis companies, and six of our men followed a few days afterwards for the same object.

On the 29th of June, the two men despatched by the R. M. F. Co. returned in a miserable plight; they had proceeded as far as Laramie's fork, at the foot of the Black hills, and were robbed by a party of Crow Indians, of their horses; after which they retraced their steps to camp, and suffered extremely for want of provisions, or from cold, rain, and fatigue. Throughout the month of June, scarcely a day passed without either rain, hail, or snow, and during the last three days of the month, a snow storm continued without intermission, the whole time, night and day; but disappeared from the earth a few hours after the sun reappeared.

On the third of July, one of our men who was sent in quest of the St. Louis companies returned, and reported that William Sublett, at the head of one hundred men, was now on his way here. This numerous company was composed of fifty hired men; a party of twenty-two men, detached from Ghant's company; a party of thirteen men from the Rio del Norte, and a Mr. Wythe with ten or twelve followers, who was on some secret expedition to the mouth of the Oregon, or Columbia River. We learned that Mr. Fitzpatrick left the company at the Red Hills, with two horses, and set out to reach us, in advance of Sublett; but had not since been heard of. Two or three nights before our express reached them, their camp was fired upon by a party of unknown Indians, but no one injured. Several horses were stolen, however; from Sublett, our express could learn nothing of Fontenelle; and determined to proceed on until they should meet him, but the day after their departure from Sublett's Camp, they were charged upon by a party of mounted Indians, who compelled them to return.

On the 8th Sublett arrived, and halted in the middle of the hole, with the R. M. F. Co., for whom he brought one hundred mules, laden with merchandise. The same evening Mr. Thos. Fitzpatrick, to our great joy, came into camp, though in a most pitiable condition. It appears that this traveller, on his way to Pierre's Hole, came suddenly upon a large Village of Indians, who mounted their horses and immediately gave chase; however, he had fortunately taken the precaution to furnish himself with two horses, previous to his departure from camp, one of which had the reputation of being fleet. This last he led by the halter, ever saddled, and bridled, as a resource in case he should be compelled to seek safety by flight. So soon as he found himself discovered and pursued, he sprang upon his favorite horse, and fled, directing his course towards the mountains, which were about three miles distant. When he reached the mountains, the Indians were so far behind, that he hoped to elude them by concealment, and immediately placed his horse in a thicket, and sought a crevice in the rocks, where he concealed himself. In a few moments the blood hounds came up, and soon discovered his horse; from his place of concealment he saw them searching every nook and crevice, for him, and the search was not discontinued, until the next step would have placed him before the eyes of a blood thirsty set of wretches, whose clemency in the first instance, is yet to be recorded. Fortunately for him, the search was abandoned, and the Indians returned to camp, at the same time he chose a point, whence he could discover any passing object, in the plain beneath him; and determined to remain, until the company should pass, and join them at that time. At the expiration of three days, he discovered six men, passing in the valley, and immediately descended the mountain to join them, but ere he could effect this, a party of Indians appeared from another quarter, and gave chase to the six men, who wheeled and fled; in the meantime, he fled back to his place of refuge. At length he became confident, that the company had passed him without his knowledge, and set out for Pierre's Hole in the night; his moccasins became worn out, and he was forced to make others of his hat, he likewise lost his powder in swimming a river, and suffered from the combined effects of hunger, cold, and fatigue, until he was reduced to a mere skeleton, and could scarcely be recognized when he finally reached camp. He informs us, that the Indians were doubtless a band of Grosvents of the prairie, who passed from the Missouri to the head of the Arkansas three years ago, and were now on their return to their own country. They are the same Indians who encamped with Smith, Sublett and Jackson, on the Arkansas last summer, and there buried their hatchets and animosity together. But it appears from their proceedings this far, that they have raised both since.

Chapter XXVII

On the 17th a party of trappers, of the Rocky Mountain Fur Company, having received supplies for the fall hunt, left the company, and passed ten miles up the valley, intending to cross on to Lewis River, near the mouth of Salt River. The following morning they discovered a party of strange Indians near the margin of the stream, some distance above them, and several of the men immediately departed to ascertain who they were. As they approached, the chief advanced to meet them, armed with nothing but the calumet of peace; but he was recognized to be a Grosventre and in a twinkling was sent to eternity. At the same time the Indians, who perhaps numbered fifty men, besides women and children, entered a grove of cottonwood trees, and without loss of time proceeded to make a breastwork, or pen of trees impenetrable to balls. In the mean time an express was despatched to inform us, and in a few minutes the plains were covered with whites, and friendly Indians, rushing to the field of battle. On their arrival, however, the enemy had completed an impenetrable fort, fifty feet square, within which they had fastened their horses. A general fire was immediately opened upon the fort, and was warmly kept up on both sides until dark. In the mean time a plan was formed by the whites to burn them up in their fort, and quantities of dry wood and brush were collected for that purpose; but the Indians on our side objected to this project, on the ground that all the plunder would be lost, which they thought to appropriate to their own use. At length night came on, and the whites, who were provoked at the Indians, for not consenting to annihilate the enemy at once, departed for their respective camps; the Indians soon followed, and left such of the enemy as survived, at liberty to depart and recount their misfortunes to their friends. We lost in this engagement, two men killed, one mortally wounded, and many others either severely or slightly. The Indians on our side, lost five killed, and many wounded, some supposed to be mortally. The following morning, a large party of both whites and Indians returned to the fort. In it were the dead bodies of three Grosventre Indians, a child, twenty-four horses, and several dogs. Our Indians followed the route of the fugitives several miles, and found their baggage, which they had concealed in divers places, as well as the bodies of five more Indians, and two young women, who were yet unhurt, though their heartless captures sent them to the shades, in pursuit of their relations without remorse. Amongst the dead horses were those lost by Mr. Fitzpatrick some days since; but those stolen from Sublett about the same time, were not among the number; hence we supposed that a larger party of Indians were yet behind.

After this period we enjoyed fine weather, and nothing occurred worthy of remembrance, until the 27th. This evening five of seven men who departed for St. Louis, three days since, returned, and informed us that they were attacked yesterday, by a party of Indians in Jackson's Hole, and that two of their number, Moore and Foy, killed. The survivors saved themselves by flight, but one of them was wounded in his thigh.

On the 30th William Sublett departed on his return to St. Louis. He had been detained here much longer than he intended, owing to a wound he had received on the 18th. During the first day's march, Stevens, the person who was wounded in his thigh, several days since, died, and was interred in the southeastern extremity of Pierre's Hole. On the first of August we had a hail storm of one hour's duration. Until this period we had anxiously awaited the appearance of Provenu and Fontenelle; but they came not, and we became apprehensive that they had lost their horses on the way, and were thus prevented from reaching us, according to promise however, Dripps and Vanderburgh resolved to move over to Green River, and learn if possible something definite. We set out on the 2d and reached the head of Pierre's Hole on the 3d. On the 4th we crossed the mountain, and descended into a large prairie valley, called Jackson's Big Hole. It lies due east of the Trois Tetons, and is watered by Lewis River, which leaves the valley through a deep cut in the mountains, impassable for pack horses; hence trappers have to cross the mountains to Pierre's Hole, in order to avoid greater obstacles, which present themselves at any other pass. The waters of this river, in the head of the Hole, expand into a lake of considerable magnitude, which I believe is identical with one attached to the Big Horn River, on the maps of the United States, for I have never heard of any lake on the sources of that river, although our trappers have explored every spring source of it. This lake is called the Teton Lake, from the mountain that overlooks it. The river flows through the valley in a southwest direction, and near the lower end of the hole, a large branch from the southeast falls into it. Those streams are bordered by aspen and cottonwood trees, and groves of cedars, in some parts of the valley. The Hole is surrounded by lofty mountains, and receives its name from one of the firm of Smith, Sublett and Jackson.

We crossed Lewis River at a well known ford, where its waters are separated by several Islands, and are expanded to the distance of several hundred yards; but are fordable at this season for pack horses, if led carefully over, following the bars or shallow places. In the evening we halted on a spring, four miles east of Lewis River, after marching twenty-two miles. On the 5th we passed six or eight miles southeast, and halted on the margin of the stream, flowing from that direction. During our march, some of the hunters saw the bones of two men, supposed to be those killed from a party of seven, in the latter part of July. On the sixth we entered a dark defile, and followed a zig-zag trail along the almost perpendicular side of the mountain, scarcely leaving space in many places for the feet of our horses; we all dismounted, and led our animals over the most dangerous places, but notwithstanding this precaution, three of them lost their footing, and were precipitated sixty or seventy feet into the river below; two were but slightly injured, having fortunately fallen upon their loads, which preserved them from death; but the other was instantly killed. At length we came out into an open valley after a march of fifteen miles, and halted in its eastern extremity. This small valley is called Jackson's Little Hole, in contradistinction to its neighbor, which we left yesterday. It was covered with herds of buffalo, numbers of which fell before our rifles, and supplied us with fresh meat, an article we had not possessed since we came into Pierre's Hole. We saw several encampments of a large village of Indians, who had been in the valley five or six days since. They were doubtless Grosventres of the prairie, and were prevented from passing by way of Pierre's Hole, most likely, by the reception met with by a small party, who reached that Hole in advance of the main village.

On the 7th we ascended a high abrupt hill, covered with dense groves of aspen trees, and came in view of a vast plain, gently descending eastward to Green River, which flows through it southeastward. The plain was literally covered with buffalo, numbers of which we killed, and halted at a spring on the summit of the hill. On the 8th we descended the plain to a stream flowing into Green River, and halted on its margin; during the day we discovered a party of horsemen several miles to the northward, who were supposed by some, to be our long expected company, and by others were believed to be the Grosventres, who we all knew could not be far in advance of us.

To our great joy, however, they proved to be the former, headed by our old friend Fontenelle, who had passed from St. Louis to the mouth of the Yellowstone River in a steamboat, and thence with pack horses to this place. He had about fifty men, and three times that number of horses, and was aided by Mr. Provean in conducting the expedition. He fell in with the Grosventres two days since, on Green River and although they numbered five or six hundred warriors, want of ammunition prevented them from making an attack upon him; they denied having any knowledge of whites in this part of the country, notwithstanding we had given them sufficient cause to remember us, at least for a few days. He likewise saw a company of one hundred and twenty men, with twenty covered wagons, and numbers of pack horses, led by one Captain Bonyville from New York, who was at this time constructing a fort on Green River, a few miles below us.

Chapter XXVIII

On the 12th all arrangements, for the journey being completed, Mr. Fontenelle departed with thirty men, and the furs we had collected during the past year, for Fort Union at the Yellow Stone; at the same time Messrs. Vanderburgh and Dripps, who were now jointly acting for the American Fur Co., departed at the head of about ten men, intending to hunt on the source of the Missouri. We reached a spring, on the summit of the hill, east of Jackson's Little Hole, in the evening; and halted for the night. On the 14th we passed through the Narrows, between Jackson's Holes; and avoided some of the difficulties we met with on our previous passage, by crossing the river, several times. In the evening we halted for the night near the remains of two men, who were killed in July last. These we collected, and deposited in a small stream, that discharged itself into a fork of Lewis river; that flows from Jackson's Little Hole.
On the 16th we reached the head of Pierre's Hole, and found the bones of several Indians, who were supposed to have been killed during the battle in July last; and were transported here by their relations, though several miles from the battle field. Three days after we reached Henrie's Fork amid clouds of dust which rose from our horses' feet, and filled our eyes. The plains were covered with buffalo, in all directions, far as we could discern them.
On the 20th I departed with two others, with orders to seek the Flatheads, and induce them to meet the company in Horse prairie, if possible, in eight days from this time. Our leaders intended to cache their goods at that place, and wished to meet the Indians, for the purpose of trading with them. Our company continued onward a north course, whilst we passed north of the sand mountain, and bore a trifle south of west, in the direction of Cota's defile. We reached Kamas creek at sunset, after a march of forty-five miles, during which we suffered extremely, owing to want of water, on the route; but allayed our parching thirst when we arrived; ate a hearty supper of dry meat, hobbled our fatigued horses, and slept in a thicket until sunrise. Next day proceeded on thirty-five miles, to Cota's creek, and halted until dark. During our march we saw traces of horsemen, who had passed by recently. At dusk we passed two miles up the defile, and halted in the logs, near the margin of the creek. On the 22nd we mounted our horses, at day break, and passed the narrows into a rolling plain, where we found several encampments made by the Flat heads twenty days since. At noon, we halted to bait our horses, and demolished a few pounds of dried meat, ourselves. At the expiration of two hours, we again departed; and proceeded down the plain, until

near midnight, halting at length near the margin of a small stream. During the night our slumbers were disturbed by the bellowing of a herd of bulls, near us; and by the howling of a multitude of wolves, prowling about the buffalo. We were approached, by a formidable grizzly bear, who slowly walked off, however, after we had made some bustle about our beds. We made during the day and night, about fifty miles.

On the 23d we arose in the morning, and found ourselves in the valley of the east fork of Salmon river. There were large herds of buffalo slowly moving up the valley, which led us to believe, that the Indians were not far below us. One of their encampments appeared to have been evacuated, but five or six days since; and was at this time a rendezvous for wolves, ravens, and magpies. We likewise saw numbers of salmon, forcing their way up the small streams, in this valley - many had so worn out their fins, that they could with difficulty avoid us when we endeavored to catch them, in our hands. With clubs and stones, we killed several of them, with which we regaled ourselves at noon, and my companions, amused themselves, whilst our horses were feeding, by adding to the numberless carcasses scattered along the shore, that had been taken and thrown away by the Indians. We passed through this valley, and halted some time after dark at the mouth of a stream from the south, after travelling forty miles.

On the 24th we passed between two high rocky points jutting into the river, and came out into an open plain two miles wide. Near the entrance, is a bed of stone, which is frequently used as a substitute for soap. It is but little harder than chalk, of the same color, and when manufactured into pipes, and burnt, becomes a fine glossy jet color, and equally hard as stoneware. In this plain we discovered an encampment that appeared to have been made so recently, that we were confident of finding the Indians before night; however, we followed the trail to the forks of Salmon River, passing several other encampments, which were now occupied by bears, wolves, ravens and magpies, which were preying upon the yet undevoured particles of dried meat, and fragments of skins scattered around them. At dark we halted near one of these encampments in the forks of Salmon River, after riding about forty miles. In the night we were serenaded by the growling of bears and wolves, quarelling for the half-picked bones about them.

Chapter XXIX

On the 25th we continued down Salmon River to a high abrupt plain, jutting down on the east side, which leaves a narrow trail along the brink of the river for several hundred yards, over-hung by a frowning precipice some hundred feet high. Through this we passed, and came into a small prairie, decked with huge fragments of rocks, trees, and willows. On the neighboring hills, we discovered a colt that had been left by the Indians, and likewise an encampment on the margin of the river that had evidently been left yesterday; we followed the trail over ranges of prairie hills, and finally found an encampment that had been left this morning, the Indians having crossed the mountains in the direction of Bitter-root River.

Having already exceeded the time alloted us by our leaders, and being aware that they would not wait more than a day beyond the time for us; I was forced to abandon the pursuit, or risk not seeing the company, until the expiration of the fall hunt which would subject me to complaint, as well as danger; and every hour's ride being two from the place of rendezvous, I turned my horse up a small stream, and followed it eight miles into the mountains that separate the valley of Salmon River from the Big Hole. During this jaunt, we killed a grey wolf which was fat, and made us a tolerable supper; we likewise wounded a grizly bear, but in his rage, he broke down bushes and saplings with such ease, that we concluded that it would be imprudent to meddle with him any more. We made about twenty-eight miles today, including deviations.

On the 26th we started at sunrise, and reached the head of a ravine, in the opposite side of the mountains, at sunset; after a toilsome and continual march of five or six miles, including necessary deviations from our general course. The distance attained will be proof enough of the existence of obstacles in this day's march, which was one of the most fatigueing I ever attempted. The sides of the mountains were very steep, and were covered with green or fallen pines, of which the latter were so interlocked with each other, and so numerous, that we were continually forced to leap our horses over them, and were frequently compelled to retrace our steps and seek some other passage. Here, an avalanche of huge rocks, trees, and snows had been precipitated from the summit of the mountains, and the sharp fragments left in the route, if slightly disturbed, would immediately resume their headlong course downward, and presented a barrier not only impassable for horses, but even for men. From this we turned, and sought to wedge our way through the pines in another direction, but suddenly came to the brink of some frightful ravine several hundred feet deep, but so narrow that a mountain goat would over-leap it without hesitation. Here we again turned, and followed the sharp edge of a very narrow ridge, between two dark profound caverns, which yawned in immeasurable depth and obscurity, almost beneath our feet on either side. Continuing our progress, we at length reached a small cove at the head of a ravine above the regions of pine, which was covered with banks of snow, and was nearly surrounded by a naked wall of rock, which forms the base of the huge pyramids that constitute in general the summits of the Rocky Mountains.

With great difficulty we succeeded in gaining the top of the wall between two peaks, and halted beside a vast bank of snow, from which little rills were trickling down either side of the mountains, that fall, both into the sources of the Missouri and Columbia. From this height we surveyed with pleasure, the apparently level prairies and bottoms bordering Salmon River on the one side, and the more extensive and fertile valley of Wisdom River on the other. After refreshing ourselves by a cool draught from a rivulet, which formed a reservoir a few feet from its source, we commenced our descent, which was by far more rapid and dangerous than our ascent, though infinitely less difficult. At dark we reached a cove in the upper region of pines, and gladly threw ourselves down to sleep, overcome by fatigue, having walked and led our horses the whole time, since we set out in the morning.

On the 27th we followed the ravine to a small stream, which flowed several miles with uncontrolable fury, but at length reached a point where the barriers on either side of the ravine expanded, leaving room for a beautiful little lake, two or three miles in circuit, of perfect transparency, which was surrounded by gigantic pines. From this point we continued six or seven miles and reached the open prairie of the Big Hole. During our march we killed a fine black-tailed deer, and saw the trail and an encampment of the R. M. F. company, who had passed through this valley eight or ten days since; in the afternoon we continued fifteen miles up the Hole, killed a white-tailed fawn, and halted for the night in a point of pines.

On the 28th we ascertained that the company had not passed, and chose a situation whence we could discover, any passing object in the southern extremity of this valley. Here we constructed a pen of dry poles, and covered it with branches of the balsam fir, to shelter us from storms, as well as the missiles of Indians, in case of attack, being determined to await the arrival of the company, at this place. We ate to day the small portion we had saved of the buck, and nearly finished the fawn. In the afternoon, it commenced snowing, and continued all night; the following day it snowed without intermission until we lay down to sleep. On the morning of the 30th we arose, and found the prairies covered with snow to the depth of one foot; though the storm had abated, however, the plains are so warm, that it must rapidly disappear.

On the 31st we saddled our horses, and passed two miles across the valley in quest of food, having had nothing to eat, save part of a famished wolf since yesterday morning. The snow disappeared from the plains at noon, and discovered to us traces of buffalo, which we followed into the hills on the east side of the Hole. We found the herd grazing in a narrow bottom; they were so unusually wild, however that we succeeded only in stopping a bull by one of our balls, whilst the other disappeared instantaneously. In the mean time we approached, and opened fire upon the wounded one, but night overtook us and we were obliged to leave him on his legs, after firing at him ten or twelve times. We retired supperless to a neighboring thicket, and passed the night.

September first, early in the morning we departed, hungry as bears, in the direction of the bull we wounded and left last evening. As we approached, the presence of thirty or forty wolves, proved to us, that some of our balls had been well directed; yet we could not find meat enough for breakfast, that was not torn or mangled by them. However our appetites were so well sharpened, that we were not long in cooking some half picked bones, which were quickly fastened to our saddle cords, preparatory to going in quest of firewood. In the mean time the wolves, and the multitudes of ravens, remained a few yards off, politely waiting for us to serve ourselves; hinting, however, by an occasional growl, or scream, for us to be as expeditious as possible. As soon as we departed, they simultaneously sprang or flew to the carcase, with such intimacy, that ravens were seen picking at a bone, in the mouth of a wolf.

Immediately after our departure, three men entered the valley from the eastward, and charged furiously toward us, but as they came from a point we expected the company, we rightly conjectured that they were hunters, in advance of camp. In a few moments they came up, and before we had made our usual brief inquiries, the company appeared, and we passed with them, twelve miles, northward down the valley. Nothing had occurred in camp since our departure worth noticing.

Chapter XXX

In the two following days we travelled fifty miles, and reached the northern extremity of the Big Hole, in the same part of this valley. We saw two or three bears, antelopes and deer, and great numbers of young ducks, yet unable to fly, in the streams. On the fourth we passed into the Deer-house plains, and saw the trail, and several encampments, of the Rocky Mountain Fur Co.; but no game, save one antelope.

On the fifth, we passed twenty five mile, west of north, down this valley. In the mean time, our hunters killed three grizly bears, several goats, deer, and two buffaloes; the latter, however, is seldom found in this country; though it abounds in black and white tailed deer, elk, sheep, antelopes, and sometimes moose, and White mountain goats have been killed here.

On the sixth, we left this valley, and bore northward over a low mountain, to a small stream that flows into the Arrow-stone river; the country below us, is a succession of isolated hills, partially covered with pines, and fragments of rock, or extremely small bottoms, intersected by prairie hills. On the seventh, we traversed a low mountain, to a small stream, flowing northwestward, through an irregular plain. During the day we espied a party of horsemen, at the distance of two miles, who immediately ascended an eminence, discharged their guns in the air, and reflected the rays of the sun upon us with a mirror. Some of our party went to them, and ascertained that they were Snakes, who had been on an expedition against the Blackfeet. They had succeeded in capturing a woman, with a young child, whom they put to death; and decamped with twenty horses, which they stole the same day. On the eighth, we continued down the stream fifteen miles, to a large valley, surrounded by mountains; of which those on the north were exceedingly lofty; here we again intersected the trail of the Rocky Mountain Fur Co., and judging from the fresh appearance of their traces, that they were but a short distance before us, we immediately followed, determined to overtake them, and by this means share a part of the game, which is usually found in advance of a company, but never behind. We followed the principal stream, that flows into this valley, called Blackfoot, which flows into the Arrow-stone river, at a place called Hell-gates up into the mountains, about five miles, and halted in a small bottom, for the night.

On the ninth, we continued the pursuit twenty miles farther into the mountains. During our march we saw an encampment, that was left this morning, in which fires were yet burning.

On the fourteenth we crossed the mountains, to the waters of the Missouri, a short distance above the mouth of Dearborn's river; and encamped on a small stream, with the Rocky Mountain Fur Co. From the summit of the mountain, the country presented a vast plain, dotted by table and pointed clay bluffs; which were extremely regular and picturesque, resembling fortresses, or castles, surmounted by towers and domes, which at a distance, appeared so magnificent and perfect, that one could hardly persuade himself, that they were the productions of nature; so strongly did they resemble the works of art. - Those, who have had the pleasure of seeing the elegant and correct representation of scenery on the Missouri, in that splendid collection of paintings, CATLIN'S PICTURE GALLERY, consisting of Indian portraits, views of their Villages, Buffalo Hunts, Religious Ceremonies, Western Landscapes, etc., can form a tolerable idea, of the imposing and romantic prospects, that abound in this section of the country. This extensive plain was bounded by the horison to the north and eastward, but rugged mountains presented themselves in every other direction. The Missouri winds its way through it to the northward, towards the mighty falls, described by Lewis and Clark, in all their terrific grandeur. We found the Rocky Mountain Fur Co. like ourselves, in a starving condition. They reported that a party of Indian trappers, supposed to be Black Feet, had preceded them a few days, and consequently the country was almost destitute of game; some times they had succeeded in killing a grizly bear, or black tailed deer, which divided amongst eighty men, was but a mouthful for each; though generally they had retired to bed supperless. This had been precisely the case with ourselves, since we left the Deer-house Plains. We likewise learned, that a young man named Miller, who belonged to this Company, and who was wounded at Pierre's Hole, during the battle in July last, died a month afterward, and was interred in Cotas defile.

On the 11th, hunters were despatched in quest of provisions, and returned in the evening successful; having killed a bull, together with several deer, and antelopes. In the mean time, the trappers went in search of beaver, but generally returned with their traps, of course unsuccessful. On the 12th both companies raised camp, and proceeded together southeastward, over rugged hills, to a small stream flowing eastward, towards the Missouri. During our march, we killed several black tailed deer, which were numerous in the pines, with which the hills were covered. We continued our course next day, over the same description of country, following a road composed of several parallel trails, a few feet asunder, which was evidently much used by the Black Feet, as no other Indians pass here with lodges.

Near the trail on the summit of a hill, we saw a quantity of broken bows and arrows, together with remnants of Indian garments, which induced some of our comrades to believe that a party of Indians had been defeated here a year or two since; not withstanding, bones, which are usually found on battle fields, were not seen. Others, however, inferred that these articles had been sacrificed to the malignant Deity, after some unfortunate expedition, in which they had sustained irrepairable losses.

In the evening of this day we reached a small branch, which unites with others, and is then called Vermillion river from a bed of red earth found near it, which is used by the Indians for painting their faces and clothing. Here we remained the following day, to rest our horses; whilst some of the trappers explored several small streams, in search of beaver.

On the 15th we again continued our course, over a low spur of the mountain, to a small stream that led into a fine prairie valley, eight miles wide, and fifteen in length from north-west to south-east. The Missouri is separated from it by a range of pine covered hills. Its course is marked by a chain of lofty mountains, which extend parallel with it, on the east side, and were distant about fifteen miles from us. Several of our hunters brought in today the flesh of several deer and big horns, both of which are numerous on the hills.

On the 16th, the R. M. F. Co., together with Mr. Dripps, at the head of fifty of our men, directed their course towards the three forks of the Missouri, south-east-ward. During our progress we met a severe storm of sleet, which we were compelled to face, until we reached a suitable place to encamp.

On the 17th we arose, and found the country mantled with snow, which was still rapidly falling; however, we descended the mountain, and crossed a high hill, into the deer house plains, after a long march of twenty-five miles. The storm abated at noon, but the ground was covered with snow to the depth of several inches.

On the 18th we continued twenty miles up the valley, and saw numbers of rabbits, which were pursued in various directions by our dogs, as well as a herd of elk; yet our hunters were unable to kill anything, though the carcass of a wolf would have been acceptable at this time; having killed nothing, save one or two deer, since we separated from Dripps. The following day we reached the mountain, at the head of this valley; but saw no game save a herd of antelopes, whose vigilant sentinels baffled the efforts of our hunters to approach them; and thus we starved in view of plenty.

Chapter XXXI

On the 20th we crossed the mountain, and encamped on the Jefferson, about thirty miles below Beaver Head. Here, our hunters were partially compensated for their bad-less luck previous to this time; for they brought into camp the flesh of one bull, several elk, deer, and antelopes, upon which we feasted fully.

The next day being Friday, some of our catholic comrades conscientiously kept lent, having eaten so much the day before, as to be utterly unable to violate this custom of the church, had they even felt so disposed; they are however, by and by, not often so forcibly reminded of the propriety of compliance with religious observances, though the expediency of those rites is often illustrated in a similar manner.

In the afternoon, accompanied by a friend, I visited the grave of Frasier, the Irroquois, who was killed and buried here last fall, being desirous to ascertain what was generally believed already, namely, that his body had been stolen from the grave, robbed of its covering, and thrown into the Jefferson by the Black foot Indians. This opinion originated from the circumstance of finding the body of a man in the river last fall, and was now fully confirmed by the grave being open. After this time, we continued southward up to the Philanthropy, and killed elk, deer and antelopes; and caught some beaver, on the route. Fifteen miles below Beaver Head, is a quarry of green stone, that is semi-transparent, and easily cut with a knife. It is highly prized by the Indians, for manufacturing into pipes. It is situated in a bluff, on the west side of the river; over-looking the plain. In the vicinity of the Philanthropy, we saw several fine herds of buffalo, and our hunters reported that the plains were covered with them near Beaver Head.

On the 24th several Black Foot-Indians were seen lurking about the thickets that skirt the river, evidently watching an opportunity to kill some of our trappers, who being aware of their design, always go out in parties of several together, for mutual safety.

After this period we continued southeastward, following the course of the Philanthropy, and trapping it in our route, about twenty miles to the head of this plain, where the river flows from a narrow defile, one or two miles in length. Continuing our course through the narrows, we re-entered the valley - where the Indians with us killed a Black foot last fall - and again reached the mountain, whence the river flows, after a march of fifteen miles.

On the first of October, we left the plain and followed a zig zag course of the river fifteen miles, into the mountains; halting in the evening in a narrow bottom, scarcely large enough to contain ourselves and horses; however, beaver signs were numerous, and we remained two nights, being amply compensated for the inconvenience of our situation, by the numbers of beaver we caught during our stay.

On the 30th we left the river, and ascended the mountain eastward, with inexpressible fatigue, owing to the obstructions that lay in our route, added to the perpendicularity of the ascent; though we succeeded in reaching the summit, without accident, and encamped beside a fountain on the south side, at the base of an enormous peak, that rises majestically far above the rest, is crowned with eternal snow, and overlooks the plains of both the Jefferson and Madison rivers.

On the 4th we arose early in the morning, and found the country covered with snow, to the depth of fifteen inches. Last evening the weather was pleasant, and bade fair to continue so. We halted late, and were nearly overcome by fatigue; hence we neglected our usual precaution, constructing cabins; which otherwise would have deprived us of the laugh we enjoyed, at the expense of our comrades, who successively popped out their heads as they arose, half supported, from the snow, by which they were completely buried, and which tumbling in, reoccupied their beds, the moment they left them. The day was extremely cold, and the snow continued falling so fast, that we were forced to remain; however, we prepared shelters for the coming night, and kindled large fires in the pines, by which we dried our bedding, and passed the day. On the 5th the storm had abated, though the atmosphere was still cloudy and cool; however, we descended the mountain, following a spring source until it increased to a large creek, having a rapid and noisy current. In the evening it recommenced snowing, and continued all night and the following day, without intermission.

On the 7th we raised camp, though the snow was still falling very fast, and the company crossed a low spur of the mountain, in a northeast direction, fifteen miles to a parallel stream. In the mountain I, with several others, in quest of buffalo continued our course eastward ten miles, to the junction of this stream, with the Madison river. This branch of the Missouri is here eighty yards wide, quite shallow, and its bed is composed of smooth round rocks, of a black color. It commands a narrow valley, terminated on either side by abrupt and lofty mountains, through which it flows to the northward. Its borders were decked with a few black willows, of an inferior growth, which appeared to be out of place in their present situation. There are however several small streams flowing into it, whose borders are covered with aspen and pine trees, or thickets of common willows. After we separated from the company this morning, the storm increased so much that we could discover nothing, and with difficulty kept our course; but the cutting winds became less tedious, as we approached the river, and finally abated; in the meantime we discovered a herd of buffalo, lying in a ravine sheltered from the storm, one of which we killed and went to camp. On the 8th the storm continued with fury all the day, yet regardless of its severity, we raised camp and passed over to the mouth of the creek, that we left yesterday; when we sheltered ourselves in a grove of dead aspen trees, which supplied us with an abundance of fuel. The snow is now more than a foot deep, in the bottoms bordering the river.

On the 6th our long absent friend, the sun, reappeared with such lustre, that one, without the gift of prophecy might have foretold, the rapid annihilation of the snow, which followed; leaving the country partially inundated with water. During the day, the Rocky Mountain Fur Company arrived from the three forks of the Missouri and encamped near us; they separated from Mr. Dripps at the forks, who continued up the Jefferson; whilst they trapped the Gallatin, and crossed to the Madison, a few miles below us. They had caught but few beaver, and were several times alarmed by parties of Indians, who were lurking about them, but as yet no person had been injured.

On the 20th the weather was disagreeable, and the prairie wet and muddy, which prevented either company from moving, though both were anxious to proceed. It was passed however, in the various amusements, incident to such a suspension of active operations; in which card playing was the principal; and, as if to illustrate the various subjects of conversation, and give emphatic form to particular photographs, the stentorian voices of the hardy hunters, were occasionally heard, practicing that fashionable folly and crime, profane swearing.

Chapter XXXII

On the 11th the Rocky Mountain Fur Co. raised camp, and departed southward up the river, to accomplish their design of trapping its sources, before proceeding to winter quarters. Though desirous to imitate their example, and be moving, we were yet compelled to remain quiet, and pass this day, as we had the preceding one, in inactivity; as some of our absent trappers had not yet returned.

Oct. 12th. - This morning we raised camp, passed about fifteen miles down the river, and encamped on its margin. It here passes through a narrow valley, flanked on either side by a bold bank fifty or sixty feet in height; from the top of the bluff, however, a gently irregular plain is seen, extending fifteen or twenty miles, in a northeast direction, nearly ten miles in width and bounded on either side by lofty snow covered mountains; through which its channel, a deep canal with perpendicular rocky walls of considerable height, winds its devious way - Near our encampment we discovered a herd of buffalo, and killed five of them. On the succeeding day we travelled over the plains to the mountains, which we likewise crossed at a very low pass, and halted on a small fork, that flows through a range of barren hills, and discharges its waters into the Philanthropy. Our course was north of west, and we made about eighteen miles.

On the 14th we descended from the hills, and encamped near this run, eight miles below the narrows, on a small plain, surrounded by the most imposing and romantic scenery. During our march we had an alarm of Indians from some of our hunters; and myself and others went to ascertain the truth. We proceeded, however, but a short distance when we found the remains of a cow, just butchered, and evidently abandoned in haste, which satisfied us that the butchers had fled for safety or assistance. We returned and reported the discovery to our partizan. In the mean time a rumor was current that a party would go and ascertain more of the matter, after we should encamp. Not doubting that it originated with our leader, previous to unsaddling, I went to him, and inquired if he thought it necessary for some of us to go. "No," said he, "for this reason; if there are many of them, and they are enemies, we shall see them soon enough; but on the contrary if they are but few, they are already far beyond our reach, in the neighboring mountains." I left him without making any reply, and turned out my horse; but observed him soon after in the act of re-saddling his own, which excited my curiosity to ascertain his intentions. I therefore approached him, and was informed that he had again considered the matter, and thought it best for some few of us to go, and gain, if possible, more positive information; as the trappers could not be pursuaded to hunt when danger was apparent.

Accordingly we equipped ourselves, and sallied out of camp one after another, where we collected to the number of seven, a short distance from it. We proceeded up the river about three miles, and found a fire yet burning, near a cow evidently killed but a short time previous, and also perceived traces of

Indians following a buffalo trail up along the margin of the river. The neighboring hills were covered with vast herds of these animals that appeared to be quite unalarmed, and from these favorable appearances, we were confident there were not more than seven or eight Indians in the party. We continued on about three miles further, directing our course towards the only dense grove of timber on this part of the river, where we were certain of finding them unless they had fled to the mountains. About fifty yards from the river, we crossed a deep gully through which a part of its current flows, during the spring tides, and were carefully scrutinizing the grove, on which every eye was fixed in eager curiosity, watching each wavering twig and rustling bough, to catch a glimpse of some skulking savage. Suddenly the lightning and thunder of at least twenty fusils burst upon our astonished senses from the gully, and awoke us to a startling consciousness of imminent danger, magnified beyond conception, by the almost magical appearance of more than one hundred warriors, erect in uncompromising enmity - both before and on either side of us, at the terrifying distance (since measured) of thirty steps. Imagination cannot paint the horrid sublimity of the scene. A thousand brilliances reflected from their guns as they were quickly thrown into various positions, either to load or fire, succeeded the first volley, which was followed by a rapid succession of shots, and the leaden messengers of death, whistled in our ears as they passed in unwelcome proximity. At that instant I saw three of our comrades flying, like arrows, from the place of murder. The horse of our partisan was shot dead under him, but with unexampled firmness, he stepped calmly from the lifeless animal, presented his gun at the advancing foe, and exclaimed "boys don't run;" at the same moment the wounded horse of a Frenchman threw his rider, and broke away towards camp. The yells of these infernal fiends filled the air, and death appeared inevitable, when I was aroused to energy by observing about twenty Indians advancing, to close the already narrow passage, between the two lines of warriors. Dashing my spurs rowel deep into the flank of my noble steed, at a single bound he cleared the ditch, but before he reached the ground, I was struck in the left shoulder by a ball, which nearly threw me off; by a desperate effort, however, I regained my upright position, and fled. A friend (Mr. R. C. Nelson) crossed the gully with me, but a moment after he was called to return. Without considering the utter impossibility of rendering assistance to our devoted partisan, he wheeled, but at the same instant his horse was severely wounded by two balls through the neck, which compelled him to fly; he yet kept his eye for some moments on our friend, who seeing himself surrounded, without the possibility of escape, levelled his gun and shot down the foremost of his foes. The Indians immediately fired a volley upon him - he fell - they uttered a loud and shrill yell of exultation, and the noble spirit of a good and a brave man had passed away forever.

Thus fell Wm. Henry Vanderburgh, a gentleman born in Indiana, educated at West Point in the Military Academy, and, at the time he perished, under thirty years of age. Bold, daring and fearless, yet cautious, deliberate and prudent; uniting the apparent opposite qualities, of courage and coolness, a soldier and a

scholar, he died universally beloved and regretted by all who knew him.
The Frenchman, who was thrown from his horse, was also killed; his name was Pilou.
I had not gone above two hundred paces from the ravine, before I heard Nelson calling for me to stop. I did so until he came up exclaiming "our friend is killed! - our friend is killed! let us go and die with him." Believing that I would shortly have to undergo the dying part of the affair, without farther assistance from the Indians than I had already received, I felt little like returning, and we continued our rapid flight. The blood ran freely from my mouth and nose, and down my body and limbs; I became so faint that I reeled on my horse like a person intoxicated, and with extreme difficulty prevented myself from falling. I gave my gun to one of my comrades, the three who first fled having now joined us, and succeeded in getting to camp, where I was taken down, and soon agreeably disappointed with the cheering intelligence that my wound was not dangerous, and I would shortly be a well man. It was probed with a gun stick, by a friend who had some knowledge of practical surgery, and dressed with a salve of his own preparation, by which it healed so rapidly, that after the expiration of a month I felt no inconvenience from it.
We found our comrades in camp greatly alarmed, and so confident that they would be attacked in it, that some of them, more terrified than the rest, openly expressed a determination to flee for safety. They were however, convinced by some of the more daring and sensible, of the propriety and necessity of remaining together, to secure, by a manly defence, the property in camp as well as their own lives; that by a cowardly separation they would not only lose all their effects, and expose themselves to greater insecurity, but would ever after bear the stigma of having basely and cowardly deserted their companions in the hour of peril, when a united and manly effort was alone necessary to insure safety. The timid convinced by these cogent arguments, and all somewhat reassured, it was determined to remain together, and for greater security moved a short distance at sunset, into a point of timber, where we could defend ourselves against thrice our number. Next morning we arose, having passed a very unpleasant night, unrefreshed and haggard, but satisfied that we should escape an attack; and a proposition was made that a party should go and inter the remains of our lamented friends. But few persons could be found willing to risk the chance of finding the bodies, without falling into the same snare; consequently the design was abandoned. However, we determined to go on to the caches, (which had been made in Horse-prairie during my absence, in quest of the Flat Heads, the preceding August.) Accordingly we packed up, and passed from the south side of the river to a point of mountain between this stream and the Jefferson, when we came in view of a large smoke at Beaver Head, towards which we had directed our course.
Aware now of the vicinity of an Indian village, to that place, and having had sufficient reason for believing them enemies, consternation again seized us, and we turned our course toward a grove of cotton wood trees, on the last named river; which we reached and halted at, after a march of fifteen miles. All hands

immediately set to work, and soon constructed a strong pen of trees, large enough to contain ourselves and horses, and shelter us from the balls of our foes; which made us feel quite safe and fearless. We however kept a good look out from the trees, and guarded our horses close about camp, ready to drive them into the pen at a moment's warning, in case of the appearance of Indians. But the day passed away without incident, and the night also; yet we determined to remain in our present quarters, till we should be able to ascertain the extent of our danger, and the best means of avoiding it. To accomplish this object, some of our boldest comrades furnished themselves with our fleetest horses, and rode off in the direction of the village. - They had been but a short time absent, when they returned with the welcome intelligence, that the village was composed of about one hundred and fifty lodges of Flat Heads, Pen-d'oreilles, and others, which at once quieted all our fears, and camp again assumed its wonted bustle.

Chapter XXXIII

Soon careless groups were idly loitering on the ground in various positions; others trying to excel one another in shooting; some engaged in mending their clothes or moccasins; here one fondling a favorite horse, there another, galloping, in wild delight, over the prairie; a large band of horses quietly feeding about camp; large kettles supported over fires by "trois-pied" (three feet) and graced to overflowing with the best of meat; saddles and baggage scattered about; and to finish the description, fifty uncovered guns leaning against the fort or pen ready for use, at any moment. Such was the aspect of our camp, which was now settled; and a stranger uninformed of the late disastrous occurrences, would not have discovered that anything had happened, to mar our usual tranquility.
Next morning a party went to seek and inter the remains of our murdered friends. In the mean time, we raised camp and moved to the Indian village. I was unable to use my left arm, which I carried in a sling, yet I walked about, and felt no inconvenience from it, except when riding fast, or when my horse stumbled in travelling. There was with the Indians, a "trader" from the Hudson Bay Company, and several "Engages," from whom we learned that Dripps had passed up to the caches a few days previous. In the evening our party returned, and reported that they could find no trace of the body of Mr. Vanderburgh, but had found and buried the Frenchman - Pilou. Having ascertained that these Indians would pass that place in a few days, we promised to give them a present, if they would seek, and inter the remains of the unfortunate Vanderburgh. - We departed southward on the 18th, passed up the plain about twelve miles, and halted in a very fertile bottom on the Jefferson. Continuing our course on the succeeding day, we passed up this river about the same distance, through the Rattle Snake cliffs, and encamped on a very narrow level, at its margin. Above these cliffs the river is confined on either side, by high bald or rocky hills,

through which it meanders leaving little or no ground on its borders; some few elk and antelopes are found here, and buffalo in abundance.

Leaving this place on the 20th we crossed several forks of this stream, one of which is nearly as large as the river itself, and rises in the mountains on the east side of the Big Hole. It commands a fine little valley at its head, called by some the "Little Hole" and is separated from Horse Prairie by a bald hill. Having made about the same distance as on the preceding day, we came into the valley, at the forks, where Lewis and Clark left their canoes. Our caches were situated near this place, and we found Mr. Dripps here, awaiting our arrival. We learned from him, that nothing uncommon or serious had occurred, save the loss of a few horses, which were stolen, and camp fired upon, by a party of Blackfeet Indians, during the night of the fifth; but no person was injured, though several trappers were still out hunting. - Here we remained until the 24th, when Mr. Dripps and company set out for Snake river, where he intended to pass the winter. I also departed with two men, and a small equipment for the purpose of trading with the above named Indians. We passed about fifteen miles through Horse Prairie to the "Gates," where I found a party of them, who had left Dripps two days since. These "Gates" are a high rocky conical elevation attached to a plain jutting into the bottom on one side of the river precisely opposite to the bluff rocky termination of a plain of considerable height, on the other side, but three or four hundred yards asunder; which gives to them the appearance of formidable gates, and they were thus named by Lewis and Clark.

We remained several days with the Indians, who were actively employed in hunting, to supply themselves with meat for food, and skins for clothing, against the approach of winter. A day or two after my arrival, a small party of men belonging to Capt. Bonyville's Company, and some few lodges of Flatheads, encamped with us. This party had been out in quest of buffalo meat for the company, the remainder of which, were employed in constructing a fort, on Salmon River. One or two nights previous to their joining us, their camp was boldly entered by several Blackfeet, who were discovered by a squaw; she immediately entered her husband's lodge, and informed him of their presence. Like a true brave, he sprang forth from his lodge, gun in hand, but was shot down at its entrance. All hands immediately flew to arms, but the ever cautious enemy had already disappeared in a neighboring thicket. Nothing deserving of record had happened to this company since we saw them in August on Green River. Three days afterwards a party of twenty five trappers headed by Capt. Walker, belonging likewise to Bonnyville's expedition, arrived, and informed us that they had a skirmish with a party of Blackfeet some days since, in the Little Hole, but lost nothing except a few horses, and several rounds of powder and ball. About the same period, an express arrived from the Rocky Mountain Fur Company, stating that soon after separating from us, they fell in with a party of trappers, on the sources of the Madison, who had left Dripps on the Missouri below the three forks. They were fired upon by a party of Blackfeet, and lost one man killed, and another severely wounded; a third left them about the same time, to look for a trail, and had not been heard from since; these things occurred near the three forks before named. The Rocky Mountain Fur Company crossed from the Madison, to the head of Pierre's fork of the Jefferson, where they encountered a party of about seventy Pagans, (Blackfeet). Two of their chiefs ventured up to camp unarmed, and were permitted to go in. They expressed a wish to bury all animosity, and establish peace and amity with the whites. - They promised to meet and trade with them on Snake river the coming winter; and sent word to the Flatheads, that they should pay them a visit in the spring, and if possible exterminate their race. They stated likewise that all the Pagan chiefs had resolved in council to kill and rob the whites no more.

And at the same time they cautioned them to be on their guard against a party of more than one hundred Blood Indians, who were two days in advance of them, and might possibly "show fight." The Indians departed on the morning following, apparently much pleased with the whites, and particularly with some trifling presents they received.

Agreeable to the intelligence received from the Pagans, the Company fell in with the Blood Indians, and had a skirmish with them. The whites commenced the firing, and the Indians immediately displayed a white signal on a pole, and the firing ceased. Two of the chiefs then went into camp, and said they were sorry the whites had fired on them, as they wished to be friendly. McKenzie at Fort Union, had told them to exhibit a white flag, and the whites would permit them to come unmolested into their camps, and trade with them. They corroborated the statement of the Pagans, and said that they were now going against the Snakes. They also left the whites much pleased with the presents which they received. The R.M.F. Company arrived soon after the express, and remained with us one day; after which we all departed, and travelled over the mountains to the east fork of Salmon River, about twenty miles; and from thence about the same distance to the forks, and finally, three miles farther to Bonnyville's fortification, situated on the west bank of the river, in a grove of Cottonwood trees. This miserable establishment, consisted entirely of several log cabins, low, badly constructed, and admirably situated for besiegers only, who would be sheltered on every side, by timber, brush etc.

Chapter XXXIV

I was undeceived, at sight, respecting this "fort" which I had been informed, was to be a permanent post for trading with the Indians; but its exposed situation, and total want of pickets, proved that it was only intended for a temporary shelter for the company, during the winter.

On the seventh of November, several Indians came to us from the village we left at Beaver Head and reported, that they had halted on the spot where Vanderburgh was killed and that they succeeded in finding his bones, which the Blackfeet had thrown into the river; and had interred them, on the margin of that stream, near where he fell. After having been satisfied that their statement was correct; I made them the promised present.

In the afternoon a detachment of the R.M.F. Company, which had been to the south westward on the snake river, and its tributaries returned and informed us, that they had been down near the Walla Walla trading house of the Hudson Bay Co., l but had made a "bad hunt" owing to the scarcity of Beaver in that quarter. They saw a village of Snakes and Ponacks, amounting to about two hundred lodges on Gordiez River; which was attacked a few days before by a large party of about one hundred and fifty Blackfeet; but it seemed they were greatly deceived in the number of their foes, for no sooner had the Snakes and Ponacks sallied out in battle order, than their enemies fled into a thicket of willows. The Snakes, however, fired the prairie which rapidly spread and soon gained the willows; they, being mostly dry, quickly disappeared in the devouring element, and the Blackfeet were compelled to reappear in the open prairie, but they were so terrified that they simultaneously fled, directing their course over a barren prairie, towards the nearest point of the mountains, distant some three miles. The Snakes, mounted on their horses, followed and continued charging and firing on them, until they reached the timber on the mountain, and could no longer proceed on horse back. The Snakes then returned back to their village, scalping their fallen enemies on the way, to the number of forty men, and five women. They were not however without loss, nine of their warriors being stretched in death on the plain, and among the number, the famous Horned Chief, remarkable for his lasting friendship to the Whites. This was the individual, it will be remembered, who alone prevented a diabolical plot to murder us on Bear river, in 1831.

Possessed of a superstitious idea, that the moon was his guardian deity, this extraordinary Indian imagined that she instructed him in dreams during his sleep; and he taught his followers to believe that he never acted but in obedience to her directions, and that he could not be killed by metal. He was the owner of an uncommonly fleet bay horse, with which at one race, he has killed two deer, and but for the lack of arrows would have dispatched a third, from the same herd. He thought his favorite deity had informed him that he would invariably be successful in war, when mounted on his favorite steed, and obedient to the divine inspiration, he always rushed headlong upon his enemies without fear of death, and rendered himself so terrible to them by his prowess, that his presence alone was often sufficient to put them to flight.

At one time, meeting a small party of Blackfeet Indians traveling on foot in the open prairie, regardless of danger, and alone, he rushed upon them, with his only weapon, a spear, and killed no less than six of their number. This great warrior, scorning to take the usual trophy of victory, returned to camp and told his young men, that if they wanted hair, with which to garnish their leggins; they would find some at a given place in the prairie. Several young warriors set out instantly, and soon returned, bearing six scalps to their astonished tribe.

This intrepid hero was shot through the heart with a ball, which immediately deprived him of life. The Snakes universally believe the ball to have been made of horn, as he had induced them to think, that he could not be killed by any metal.

Some time afterwards the R. M. Fur Co., took their departure up Salmon river, intending to pass the winter in Little Salmon river valley. A detachment of forty men, under Capt. Walker, were in the mean time making preparations for their removal to Snake river, where they were to pass that inclement season. I determined to go with this party to the mouth of Blackfoot, and thence to the forks of Snake river, where Dripps intended to await the coming spring.

After some delays, we set off on our journey, and passed about forty-five miles up the narrow and irregular valley, through which the Salmon river, confined to small and uneven bottoms by the mountains, runs. The Indian trail which we followed, crossed several steep high points, almost impassable to our now feeble horses. Our course from Bonnyville's Fort, gently turned from south to south west. At the termination of this distance, we again found ourselves in the open level country, near the lower extremity of little Salmon River valley. At this point we overtook the R. M. F. Company, and passed with them slowly up to the head of this valley, a distance of thirty miles, and there again departed from them. This company resolved to pass the winter here. We passed through Day's defile, and slowly down to the termination of Day's Creek, about fifty miles. We saw several encampments of the Ponacks, who had recently passed here, in the direction of Porteneuf; at this place, Mr. Fitzpatrick of the R. M. F. Co. joined us, with one man; intending to go with us to Dripps, with whom he had some business to transact. Departing thence, we directed our course towards the lower or south-western Butte, and halted on Gordiez River, after a march of twelve miles. We saw during the day several herds of buffalo, but killed none until after we had encamped, when one of our hunters succeeded in approaching a herd of bulls, and shot a very fine one.

Next day we continued our course, and halted at a small spring in a ravine on the N. E. side of the Butte, which is the only water found at this mountain; and even it is lost in the sand before reaching the prairie. After leaving the spring, we passed east of south, twenty-five miles without finding any water; and halted at a spring, five miles west of Snake river; and seven or eight above the mouth of Porteneuf. From this place, we crossed Snake River, and encamped in the rich luxurious bottom, on the East side of this stream, December 11th. Hunters were immediately dispatched in quest of game; they returned successful, and reported, that they had heard guns firing, seen buffalo running, and discovered a large smoke on the river, about twenty miles above. Several Ponacks came to camp the next day, reported that their village was on the Porteneuf, and that they had no knowledge of any whites on the river, except a party of Norwest trappers they had seen very far down it. Shortly after an express arrived, bringing information, that four men belonging to a detachment from Bonnyville's Company, which separated from him on Green River, were killed about a month previous, near the sheep Rock; and that the remainder of the party were in winter quarters in Cache valley.

Chapter XXXV

On the 17th I set out in company with Mr. Fitzpatrick and four others, in search of Mr. Dripps. Travelling a distance of 70 miles up the river, to the forks, we fell in with some of his hunters, who escorted us to their camp, on one of the numerous islands in Lewis' river. We remained here two days, which was agreeably passed, every lodge being graced with racks which were well filled with the best of meat. All were supplied with good quarters, and appeared to want for nothing this dreary country could afford, that would contribute to their comfort or amusement. We left this place on the 22d, on our return; Mr. Fitzpatrick to join his party on Salmon River, and I, for the men and baggage we had left at the mouth of the Blackfoot. We arrived at the quarters of Capt. Walker on the 24th, and passed the next day with this gentleman very pleasantly, receiving the best treatment his - in this country necessarily limited - means would afford. During the last two days, the snow hitherto rare, had fallen to the depth of seven or eight inches. With the design to purchase a few skins of the Ponacks, who were encamped at this time on a small stream near Blackfoot River, I visited their village on the 20th, and found these miserable wretches to the number of eighty or one hundred families, half naked, and without lodges, except in one or two instances. They had formed, however, little huts of sage roots, which were yet so open and ill calculated to shield them from the extreme cold, that I could not conceive how they were able to endure such severe exposure. Warmly clad as I was, I could hardly think it possible to pass one night in such a miserable shelter without freezing, unless supplied with the means of keeping a good fire, during the whole time. They kept small ones, burning in the centre of their cabins, and groups of half starved, and almost wholly frozen women and children, were squatting in a circular form round them, ever struggling, with a feeling and energetic devotion, still nearer to approximate the element they so fervently worshipped. In almost every family might be seen several dogs crowding between the children, to share with them a portion of the animal and artificial heat, diffused equally to all who formed the surrounding ring, which is generally involved or lost from view, in the dense clouds of smoke, which custom or habit had rendered less disagreeable to them, than would be imagined.

Having succeeded in procuring a few skins - their poverty forbade my getting many - left them on my return, reflecting on their intolerable indolence, and abject condition, and its natural consequence; feeling pity for their sufferings, and yet forced to blame their entire want of industry, which might in a great measure alleviate their hardships; for a few hours employment would suffice to form a cabin of grass and branches, infinitely more comfortable than their present abodes.

Three days after, I set out with several others to rejoin Mr. Dripps, whose camp we reached on the third day; nothing worthy of record having occurred on the march. We learned here, that during our absence a man who was lost on the

Missouri, last fall, had returned to the company, having rambled about the mountains alone, for more than a month.

The first of January, 1833, or New Years day, was spent in feasting, drinking, and dancing, agreeable to the Canadian custom. In amusements such as riding, shooting, wrestling, etc. when the weather was fair, and in the diversion of card playing when the state of affairs without would not permit athletic exercises, the month of January passed away, during which, we had changed our camp three times, in order to obtain better grass for our faithful animals. The weather was generally fine, but little snow had fallen, and we usually found plenty of game near our camp - therefore time passed away not only comfortably but pleasantly. On the seventh of February, three of our trappers went up the river about twenty miles in search of beaver, when they discovered five comical Indian forts, and supposing them tenantless, they approached them without apprehension intending to pass the coming night, in one of them; when they had arrived within a few paces of them; seven or eight Indians rushed out and fired upon them. One of their horses was shot down beneath his rider, who sprang up behind one of his comrades, and they fled unharmed back to camp, which they reached the same evening. We had on the twentieth an alarm from some of our hunters, stating the appearance of a large party of horsemen, on the opposite side of the river. Some of us concluded they were the same party of Indians, who had promised the R. M. F. Co. to come and trade with them on this stream during the present winter. A party sallied out to obtain information, and soon ascertained that a large herd of elk had caused the alarm. - These animals, when frightened or startled, throw up their heads, which their long necks enable them to do quite high, and have at a distance, much of the appearance of a band of horsemen - I mention this circumstance, because when they are advancing, or retiring, at a distance of three or four miles, the most sagacious Indians are often deceived by them; and cases when horsemen are mistaken for elk are by no means uncommon. The month of February, thus far, has been very pleasant; the days are mild and serene, and seldom miss the genial radiance of the sun - the nights, however, are by no means so comfortable, and the river is frozen to a depth of two feet. The snow has in many places quite disappeared, and the returning warblers have already announced the approach of gentle spring; whom we soon expected to see, arrayed in a joyous garb of leaves and flowers. Our hunters as usual leave camp about daylight, and generally return in time for breakfast, laden with supplies of meat of various kinds, so plentiful is game in this region. Several visitors from both the R. M. F. Co., and Walker's camp, arrived about this time, and from them we learned, that five men left the former company about the first of December, on Gordiez River in quest of meat, but were never afterward heard of; and the trail of a party of Indians going in the direction they had taken, was discovered a day or two subsequent. The possibility of their having voluntarily abandoned the company, and gone to the Spanish settlements or elsewhere, is at once refuted by the fact, that they left property behind them to the amount of several hundred dollars. They were without doubt killed by the Indians; their names were Quigleg, Smith, Smith, the other two not recollected.

We remained quietly awaiting the disappearance of snow and ice, which was realized about the twenty-fourth of March. Geese and swans are now performing their migratory returns, and are continually seen flying over us; ducks are also observed in abundance. Our numerous company was now divided into two parties, one of them headed by a Spaniard named Alvaris, amounting to about forty men, departed up Henry's Fork, intending to hunt on the Yellow Stone River; and finally join us on Green River, at the expiration of the spring hunt. Mr. Dripps with the remainder, including myself, marched a short distance up Lewis River, and halted, the weather being yet so cold and wet, as to render travelling extremely uncomfortable.

Chapter XXXVI

From this time forward until the 19th of April, the weather continued raw and windy, with frequent storms of snow; yet many of our trappers were successfully employed in taking beaver. Four of them returned quite unexpectedly from an expedition to Gray's Hole this evening, considerably alarmed. They came suddenly, it appears, upon several forts in a grove of aspen trees, in that place, which were still tenanted, as ascending volumes of smoke proved to their surprise, and they immediately fled, pursued by a large body of Indians, who followed with such speed that it was for some time doubtful which party would arrive first at a narrow gorge, where the only chance of saving their lives presented itself. The certainty of death if they were overtaken, or their retreat through the pass cut off, urged them on with an energy and rapidity unknown to less pressing dangers. Even their horses seemed to comprehend the peril, and seconding with generous efforts the wishes of their riders, bore them safely on, till they reached the defile, passing through which into the open plain beyond, relieved them from further pursuit. The earth was at this time covered with snow, and from their being obliged to take a route very circuitous to the pass, owing to rocks, precipices, and other obstacles, which horsemen could not safely venture over, but which the light armed, strong limbed, and swift footed Indians easily threaded, in an almost direct course to the place, which if first reached by them, cut off the only hope of escape to the poor trappers, their danger was indeed imminent, and scarcely had they passed through the defile, before the yells of the disappointed savages, arriving at the place, proclaimed how determined had been their pursuit, and how timely their flight.
Five horsemen were seen on the 19th, on the margin of Snake river, down which they turned and fled, on finding themselves discovered. Who they were could not be ascertained, but they were supposed to be Blackfeet. - Two days afterwards, we discovered a large smoke apparently near the forks, probably proceeding from the fires of Alvaris, from whose camp four men returned on the 22nd, and reported they were still on Henry's Fork, having been prevented from

advancing further by the snow, then two feet deep on Cammas prairie; and were waiting its disappearance. Next day they returned back to their own quarters. We removed on the 24th to Gray's creek, about eight miles, and encamped in the vicinity of a herd of buffalo; several of which were killed by our hunters. In the evening a small party of trappers came in from Salt River, where they were to have remained during the hunt; they were driven to the necessity of returning, by the presence of hostile Indians. Our hunters on the succeeding day, killed a number of fine fat bulls, and as usual, we fared well.

To the reader, it may seem trifling to record the simple fact of our having heard guns fired in the cedars on Lewis river, and likewise in the direction of Gray's Hole; but to the hunter of the Rocky Mountains such an occurrence is an event of importance, and should by no means be unheeded. Surrounded by tribes of savages, whose ethicks counsel theft and murder on every occasion, and authorize treachery and cruelty without discrimination, nothing but the most watchful care, and sagacious prudence can render him even comparatively safe, in the midst of so many dangers as are constantly thrown around him, by a wicked and wily foe. True the report of fire arms may indicate the vicinity of friends, but they may much more likely herald the approach of enemies; and the most common prudence will show the importance of a careful attention to these and other alarms. The most important event in the march of a week, may be the report of a strange fusil, so uncertain are the circumstances it may produce, and so probable the vicinity of danger.

On the 26th we proceeded to Gray's Hole, twelve miles, where we remained until the 3d of May following. Scattered about the hills near our camp, I saw a great number of porous rocks, each having a cavity much like an oven, with invariably a single orifice; which in one of the largest was so small as scarcely to admit a wolf, yet the area within was sufficiently large to have contained several men. These rocks consist of a very coarse sand stone, extremely hard and nearly round, the cavities within were also generally circular, and what was very singular, though rocks of this species were numerous and of various sizes, yet each one had a similar cavern. Different species of rocks are found here in abundance, such as granite, limestone, etc., but none of these were of the form, or had the peculiarity which characterized the sand stone formation. Could the waves of the ocean, which was evidently once here, have washed these fragments of stone into such regular form, and by continued attrition have worn out those remarkable cavities? - their similitude refuses the probability of the suggestion. - Could their nucleii have been of perishable material, and in the lapse of ages, by decomposition have left those singular cavities? - the uniformity of their appearance, and the fact that each rock of the kind had one and but one aperture, will not allow the conclusion. Were they formed by the industry and ingenuity of man? - if so, to what purpose was an amount of labor expended at once so vast and so difficult? It must remain an enigma for the present. The investigations of philosophers may hereafter elucidate their origin and uses, now a mystery.

Chapter XXXVII

We changed our encampment several times further south, and finally proceeded eastward about fifteen miles and halted on a small stream, which passes here through a beautiful valley, skirted as usual by lofty hills, covered with evergreen pine. A young man by the name of Benjamin Hardister, who came out last summer with Bonnyville, but had left him and taken refuge in our camp in the winter, died on the evening of the 8th, of some complaint, the germ of which he had no doubt brought with him from the United States. With the assistance of a man behind him on the same horse, he rode eight miles during the day previous to his decease. We buried him as decently as circumstances would permit the next day, "and left him alone in his glory." On the 10th, we crossed the mountain with difficulty, in consequence of the narrow and irregular condition of the paths formed by buffalo, passing sometimes along the uneven bottom of the ravines, sometimes up the broken and steep declivity of their sides, winding often among fragments of rock, and occasionally through the almost impassible pine forests, that cover the middle region of the Rocky Mountains; and after a very tiresome march of twenty miles, found ourselves on Salt river, in a fine valley about fifteen miles long, by four broad, and surrounded by high mountains, whose bases are covered by dense forests of pine and aspen. The river runs through it in nearly a north direction, and several small creeks with willow and aspen timbered margins, flow into it from the mountain.

The valley is level, having but little sage, is covered with short grass, like all other timberless plains, and is entirely free from those little holes, burrowed by badgers, often found in other valleys, which are highly dangerous to equestrians, and frequently cause serious accidents to those, Indians and others, who kill deer or buffalo by running them on horseback. Instances of valuable horses crippled by stepping in these holes while running, are of frequent occurrence in plains where these holes are numerous. We found several bands of buffalo here, and had the good fortune to kill ten or twelve. - On the 14th several of us went up this river in quest of salt.

From the head of this valley, twelve miles from camp, we proceeded three miles through a range of hills, and came to the valley of the Boiling Kettles, already described; passing up a small branch that empties into Salt river on the same side, and a short distance above the Boiling Kettles, to its head, we found several low wet places where salt was found by elutriation in considerable quantities; in one, particularly, a layer of cubic and pyramidical chrystals seven inches in thickness, found above a black, stinking, miry substance several rods in extent, furnished us with abundance. The salt found in the country is, however, more commonly attached to stones, in the bottoms of dried up pools, like ice, and requires a hard blow, in most cases, to separate them. Breaking from the strata as much salt as we could conveniently carry, we collected the fragments and put them into bags, which we lashed behind our saddles, and sallied out into the prairie on our return to camp. We visited several springs situated on the side of a bald hill, about half a mile from the Kettles. Extending several yards around these springs, the rocky cement, as well as the earth, is hollow, and the noise of our footsteps, increasing as we advanced, at length redoubled to that degree that some of my comrades refused to approach the several holes and caves found near the Kettles, which are much smaller than those in the plain, being in no instance more than two feet high. Like them, however, the water continually boils over from a small aperture from the top, ever depositing a slimy greenish matter which soon hardens into rock. There are likewise many cavities at the basis of the Kettles, several inches in diameter, from which the boiling water constantly exudes. The surplus water proceeding both from the Kettles and cavities above mentioned, flows down a plain several feet, till it empties into a deep pool three or four rods in diameter; and the outlet of this one becomes the inlet of another, two or three paces distant, of about the same magnitude. The water in both pools is of a bright yellow color, and vapors, disagreeably odorate, are continually emanating from it. Large quantities of sulphur have been deposited on the plain through which it passes, having a beautiful yellow appearance, which can be seen from every part of the valley; though at a distance it seems white. We remained about the springs some time, and set out for camp, where we arrived shortly after dark.

We passed down the river about four miles on the 15th, and encamped on its border. Snow had fallen during the past night to the depth of several inches, but disappeared about noon to-day. On the 10th, a party of us went up to the Boiling Kettles, to procure buffalo meat; we found the valley quite covered with them, but killed a few bulls only; however, the cows are poor and in most cases inferior to the bulls at this season of the year. We saw several bears, but they are not now eatable, except in case of absolute necessity. On our way back we halted for the night, shortly after dark, in the narrows, near a mound, having precisely the shape and appearance of a vastly large haystack.

The ranges of mountains which nearly surround the valley of the springs, as they follow the course of the river down on each side, become greatly approximated to each other, and confining that stream to very small borders and compressed breadth, constitutes what is here termed the narrows; and again expanding or receding from each other, form Salt river valley.

At day-break, the next morning, we set out for camp, passed from the narrows into the plain, and down it several miles, when we discovered eight or ten objects at a distance having the appearance of elk or horsemen; proceeding on our course until we met them, they proved to be a party of trappers from Bonnyville's Company. They informed us that their camp was in the hills on a small stream a mile or so from the river; that they left Walker on Bear river, and came from the head of Black-foot to that of the stream they are now on. On the following day they moved down and encamped with us.

On the 19th we raised camp, and passed through this valley northward, and halted on Lewis river, a few hundred yards above the junction of Salt river with it, after a march of eight miles. This river is confined, a short distance above us, by formidable walls or bluff mountains, forming in some places very high and perpendicular banks, impassable for even those sure footed animals, mules; hence travellers are compelled to cross the mountains into Pierre's Hole, and then again to cross them to Jackson's Hole; when if it were feasible to pass up the river direct, two thirds of the distance we are forced to go, and the fatigue of crossing two mountains, might be avoided.

A narrow valley extended a short distance below our camp, through which the river runs in a north-west direction. The weather was, at this time, cloudy, with some rain, and the shelter of the little cabins, constructed with our blankets was found quite agreeable. About four miles below our camp, we forded the river, on the succeeding day, and proceeding about six miles further down, we halted at the mouth of a small creek, in the neighborhood of which we found wild onions in abundance, and also a species of lettuce in great plenty - weather continuing wet and disagreeable, time passed along rather heavily.

We passed six miles down on the 21st, to a fine valley about ten miles long, in which the river gradually turned to the westward, and at the head of which we rested for the night. Continuing our journey next morning, we went down to the other extremity of the valley, and again halted, having travelled in a direction nearly west, with weather still hazy and uncomfortable.

Chapter XXXVIII

On the twenty-third, we ascended the point of a hill which juts in to the river at the lower end of the valley, and then passing over an irregular plain, northwest, a short distance, we reached a small stream, which we followed up into the mountains nine miles, and encamped. Our road, in many places, was almost impassable, in consequence of the thick dense growth of aspen, through which we were obliged to force our way, to the no small detriment of our clothing, and in great danger of losing our eyes, or being at least severely bruised by the numerous branches, which were continually flying back from a strained position, caused by the pack horses forcing themselves through; and which frequently coming against us with no gentle force, in their efforts to regain their natural position, seemed determined to give us a practical lesson on the elasticity of that species of timber. By no means greatful for the instruction thus forcibly imparted, we were heartily rejoiced when the task was ended, and ourselves at liberty to retire and seek repose; fully impressed with the conviction that though to spare the rod, in many instances might spoil the child; yet its too free application on an occasion like the present, would hurt our feelings extremely, without being productive of any beneficial result.

The sun on the morning of the twenty-fourth arose clear and pleasant, and with the prospect of a fair day before us, we again started on our pilgrimage. We ascended a fork and crossed the mountains, when we arrived at the head of a stream flowing into Pierre's Holes; which we followed down into the plain, four miles; there leaving it, we passed over to the stream which marks the pass to

Jackson's Hole, about two miles, and halted, very much fatigued, about one mile above the battle ground of last summer.

Next day, with one companion, I returned to Lewis river, in search of several horses which we lost while crossing the mountains the day previous. Again we passed over it, and searched every bottom on the several small forks in the mountain without success, when we went down to Lewis river, and finally found them quietly feeding in a ravine, to which they had strayed near the prairie. We succeeded in catching them and set off on our return at a round pace, which brought us to camp, some time in the evening. On the succeeding morning, in company with a friend or two, I visited the battle ground which was situate in a grove of aspen trees, several hundred yards in extent. The pen or fort was probably about fifty feet square, was composed of green and dry aspen timber, and though hastily, yet firmly constructed. It had sunk down in some places, however, from decay, below the height of two feet perpendicular. The beseiged had excavated holes or cavities in the earth, within the pen, sufficiently capacious for two or three persons to remain in, quite below the surface of the ground. These holes extended entirely round the pen; and we ascertained that the Indians had fired, in most cases, from small holes at the surface of the ground, beneath the pen or breast work, which circumstance (happily for them) was not observed in the smoke and confusion of the battle, or they would have been annihilated in a few moments. The attack was principally made on the north side, where at every tree, sticks were still seen piled up against the roots, from which the beseigers fought; who had likewise raised a heap of brush and logs, a few paces from the pen or fort, to nearly or quite the same height; and had the Indian allies not objected, in the hope of capturing their arms, ammunition and other equipments, it would have soon been so greatly increased and advanced toward the pen, as to have insured its destruction, if fired, with all its contents and defenders. Parties were also stationed behind trees, and clusters of willows on the other sides of the fort, which was thus entirely surrounded. The trees both within and outside of the pen, were covered with the marks of balls, or of the axes successfully employed by our comrades, to exhume and save them; lead being very valuable in these remote regions, where it is so extremely necessary, both to the purposes of defence and subsistence. Bones, of both men and animals, lay scattered about, in and around the pen, bearing evident indications of having contributed their fleshy covering, to the sustenance of wolves and ravens; who undoubtedly gratified their gastronomical propensities, after a protracted fast, for some days subsequent to the conflict.

Mr. Fitzpatrick received information from the Blackfeet, whom he encountered last fall, that the Gros-Vent's after separating from Fontenelle on Green river, last August, passed over to the head of the Yellow Stone, and were there discovered and attacked by the Crows, who captured their horses, took many women and children prisoners, and killed a great portion of the men. Some few, however, of these unfortunate wretches escaped, and after enduring great hardships, reached their friends the Blackfeet, in a state of starvation, nakedness and suffering seldom met with, even among those prowling robbers who are frequently out from the villages for months together, in search of opportunities to steal horses from either whites or Indians, and hence are often necessarily exposed to cold, hunger, privation and danger, almost beyond the limits of human endurance. On our return to camp, we learned that one of our men had been severely wounded by a grizzly bear, during an excursion for buffalo, in our absence. It appeared that himself and several others discovered one of these formidable animals, near a grove of willows on the margin of a small stream. They approached and mortally wounded him; but he succeeded in crawling into the brush. Our unlucky comrade, unwilling to let the animal escape, advanced to the bushes, and was at the same instant attacked by the enraged bear, who sprang upon and threw him. His companions were so paralyzed with the fear that he would be torn to pieces, that they could render him no assistance, with the exception of one, a well known one eyed Spaniard by the name of Manuel, a famous hunter, who, quick as thought, threw his gun to his eye, and fired; fortunately the ball was well directed, and the huge beast fell lifeless beside the prostrate man, who escaped with life, but was so severely bitten in the hand, arm and thigh, as to be unfit for duty for several weeks.

Chapter XXXIX

We passed slowly over to Jackson's Hole, a distance of twenty miles, and halted, on the 31st, on a small "slough," west of Lewis river. This word is used in the mountains to designate that portion of a river separated from the main channel or current, by the intervention of an island. We passed some immense banks of snow on the mountain, but succeeded in getting over without accident; though not without difficulty; and after the fatiguing march of the day, were well drenched in the evening, by a shower of rain, which the reader may justly conclude, did not, in any degree contribute to our comfort or complacency. Though very little pleased with the aspect of the weather on the morning of June 1st, when we arose but little refreshed, and found the rain still falling at intervals, yet we raised camp, passed up the valley about fifteen miles, and halted on the east side of the river; our course having been nearly north-east. We found a large herd of buffalo in the valley, and killed several; also a large bear, which paid with his life the temerity of awaiting our approach.
Next day we left the river, and travelled eastward about four miles, to a stream some forty paces in breadth flowing into it, called "Gros Vent's Fork," from that tribe having passed here some years since on their way to the Anipahoes, with whom they are on terms of intimacy and often exchange friendly visits. This stream is quite shallow, having its bed formed of large round black rocks, over which the current dashes with noisy rapidity. Its borders are quite naked in some places, though generally a few scattering trees appear, overhanging the water. During the afternoon it rained considerably, and we found a pretty close approximation to our camp fires by no means detrimental to comfort.

On the 4th we went up this stream into the mountain eight miles, and halted at the mouth of a small fork. Our way, (it could hardly be called a road, though we were all on horseback,) was tolerable, if we except a dry red bluff, projecting into the run at high water, over which, at such times, passers are compelled to follow a narrow trail, scarcely wide enough to secure footing at the very brink of a frightful precipice, whose base is washed by the river, for several rods; but when the river is low, this may be passed below without danger. At the foot of the bluff, are the bones of many buffaloes and elk, that have been precipitated over it and killed.

June 5th we ascended the left fork five or six miles, and halted in a very small valley on the right fork of three into which the stream is subdivided. The "Trois Tetons" bear due west, consequently our course from Jackson's Hole has been directly east. This section of the country is graced with irregular clay bluffs of various colors, as red, yellow, white, etc., which give it a pleasant and variegated appearance.

We crossed over a low "divide" to a small stream on the 6th, which we followed to Green river, about ten miles distant from our last encampment, at which place we found several trappers who had been absent upwards of a month. They had caught some beaver, and met with no accident to interrupt their occupation.

With a single companion, I departed on the morning of the 7th, to ascertain if any of our long absent friends, who left us at "Pierre's Hole," with John Gray nearly a year since, had arrived at "Horse creek," the appointed place of rendezvous. Passing down the plains of Green river, twenty miles, we discovered several squaws scattered over the prairie engaged in digging roots, who informed us that a party of whites and Snakes, were now at Bonnyville's fort, a few miles below. We continued on our way down, and found at the place indicated by our informants, Capt. Walker with some of his men, also John Gray, and a small party headed by Fallen and Vanderburgh, who received an outfit from Wm. H. Vanderburgh, in Pierre's Hole last year. These different parties had made good hunts, without being molested by unfriendly Indians. One of the partisans, Fallen, went to Taos last winter for supplies, and on his return lost two Spaniards, "Engages," who were frozen to death on their horses. He also suffered greatly from cold and fatigue. One of Capt. Walker's men had been attacked by a brown bear, but escaped with a broken arm. Some fifty or sixty lodges of Snakes lay encamped about the fort, and were daily exchanging their skins and robes, for munitions, knives, ornaments, etc., with the whites, who kept a quantity of goods opened for the purpose of trading in one of the block houses, constituting a part of the fort. This establishment was doubtless intended for a permanent trading post, by its projector, who has, however, since changed his mind, and quite abandoned it. - From the circumstance of a great deal of labor having been expended in its construction, and the works shortly after their completion deserted, it is frequently called "Fort Nonsense." It is situated in a fine open plain, on a rising spot of ground, about three hundred yards from Green river on the west side, commanding a view of the plains for several miles up and down that stream. On the opposite side of the fort about two miles distant, there is a fine willowed creek, called "Horse creek," flowing parallel with Green river, and emptying into it about five miles below the fortification. The river from the fort, in one direction, is terminated by a bold hill rising to the height of several hundred feet on the opposite side of the creek, and extending in a line parallel with it. - Again on the east side of the river, an abrupt bank appears rising from the water's edge, and extending several miles above and below, till the hills, jutting in on the opposite side of the river; finally conceal it from the sight. The fort presents a square enclosure, surrounded by posts or pickets firmly set in the ground, of a foot or more in diameter, planted close to each other, and about fifteen feet in length. At two of the corners, diagonally opposite to each other, block houses of unhewn logs are so constructed and situated, as to defend the square outside of the pickets, and hinder the approach of an enemy from any quarter. The prairie in the vicinity of the fort is covered with fine grass, and the whole together seems well calculated for the security both of men and horses.

Chapter XL

On the 8th, I returned to camp, which had moved down and was now in a fertile bottom fifteen miles below the fort. Here we remained tranquilly, nothing worthy of record occurring until the evening of the 11th, when four of our trappers, who had been absent from camp some time, returned in a state of perfect nudity and most unparalleled misery. Their bodies were broiled by the heat of the sun to that degree, that the pain produced by coming in contact with our clothes was almost insupportable.
It seems that, four days previous, they prepared a raft for the purpose of crossing Lewis river; having before ascertained that it was not fordable, and when every arrangement was completed, they drove in their horses, which swam over safely, and landed on the shore opposite. Stripping themselves for greater security, they pushed off into the stream. The velocity of the current, however, capsized their raft, on which their guns, traps, saddles, blankets, beavers and clothes were fastened, and carried the whole under an immense quantity of floating drift wood, beyond the possibility of recovery; and they only saved their lives by swimming, which they with difficulty accomplished, and faint, weary, and despairing, landed on the other side, reduced by this unfortunate accident to a condition the most miserable and hopeless. - A few moments reflection, while taking a little rest, convinced them that their only chance of saving their lives, lay in endeavoring to find and reach our camp, naked and entirely destitute of arms, provisions, and other necessaries as they then were. With scarcely a ray of hope to cheer them on their dreary task, they mounted their bare backed horses, and started in quest of us. The burning heat of the sun parched their skins, and they had nothing to shield them from his powerful rays; the freezing air of the night chilled and benumbed their unprotected bodies, and they had no covering to keep off the cold; the chill storms of rain and hail pelted mercilessly on them, and

they could not escape the torture; the friction produced by riding without a saddle or any thing for a substitute chaffed off the skin, and even flesh, and without any means of remedying the misfortune, or alleviating the pain, for they were prevented from walking by the stones and sharp thorns of the prickly pair, which lacerated their feet. They were compelled, though the agony occasioned by it was intense, to continue their equestrian march, till amidst this accumulation of ills, they reached our camp; where by kind treatment, and emollient applications, their spirits were restored and their sufferings relieved. Add to the complication of woes above enumerated, the knawing pangs of hunger which the reader will infer, that they must have experienced in no slight degree, from the fact that they did not taste a morsel of food during those four ages of agony, and we have an aggregate of suffering hardly equalled in the history of human woe.

After the return of our comrades, as above related, nothing of note occurred for some time; during which we enjoyed the luxury of pleasant weather, and a fine air.

On the 25th, I departed, with some others, to meet the St. Louis Company, who were daily expected. We passed along the plain at the base of the wind mountains, a very extensive and lofty range, crossing several small creeks, which at some distance below us unite and form New Fork. While on our march we killed an antelope, and also a brown bear, which last we discovered on the open prairie, several miles from any timber, and immediately gave chase to him; but as soon as we approached to within a few paces, he wheeled and pursued those who were nearest to him; the superior fleetness of our horses, however, soon put us beyond his reach, and he again bounded off in another direction, we in turn again becoming the pursuers. We fired several shots at him, some of which, though well directed, only served to increase his fury. For a considerable length of time we were alternately pursuing or pursued, till finally after many shots had been fired at him, a ball penetrated the scull between his eyes, and simultaneously put a period to our sport, and his existence.

"At the close of the day" we retired to rest, in the open prairie, as usual, with a blanket for a bed, a saddle for a pillow, a robe for a covering, and the clear blue star-studded sky for a canopy. I slept, and gentle visions mingled with my slumbers; home, mother, sisters, brothers, passed before me in the pleasing panorama of a dream; again I seemed to tread the lovely streets of my native village, which had become a populous city. I sought and found the well remembered haunts of my childhood, and pressed in imagination, the friendly hands of the companions of my youth. The scene was too delightful to be real, and from the sweet delusions of fancy, I awoke to the unwelcome conviction of absence from home, and the certainty of comparative solitude, in the Rocky Mountains. We arose, and pursued our journey, "ere morn unbarred the gates of light," and soon after sunrise discovered the plains covered with buffalo, but so unusually wild that the report of a single gun set them all at once in motion. Few persons, even in these romantic regions, have ever witnessed so interesting a scene as was presented to our view from an eminence or high mound, on which we were fortunately situated, overlooking the plains to a great distance. Immense herds of bison were seen in every direction galloping over the prairie, like vast squadrons of cavalry performing their accustomed evolutions. Platoons in one part filing off, and in another returning to the main bodies; scattering bands moving in various courses, enveloped in clouds of dust, now lost, and now reappearing to view, in their rapid movements; detachments passing and repassing, from one point to another, at full speed; and now and then a solitary patriarch of the mountain herds, halting for a moment behind the dashing cohorts, to ascertain, if possible, the cause and extent of the danger and alarm; but soon again with instinctive impulse, hurrying to join his less fearless files; and all rushing on, till forms and numbers disappear in the dust and distance, and nothing remains visible of the long black lines but dark clouds slowly sweeping over the distant plains, which soon dissolve, and leave no trace of the tranquil thousands of buffalo, so lately grazing in peaceful quiet on the wide plains beneath our wondering, wondering vision.

When these scenes of living interest had faded into absolute obscurity, we descended from the "Butte," and continued on our way as far as the Sandy, where we halted a short time for breakfast, which having despatched, we again set off, and soon after discovered the traces of two horsemen, going in the direction of our camp. - From the circumstance of our not having before seen them, which we must have done had they been on the plains; we concluded that when we stopped for breakfast they were on the same stream a short distance above us, and had started again shortly after we halted; also that they were sent express in advance of the St. Louis Company, to announce its arrival. We, however, continued on, passed over the high barren plain which separates the waters of the Platte from those of Green river, and having descended to Sweet Water, finally halted on its margin.

Chapter XLI

On the 20th of July a young man by the name of Newell and myself, departed at the head of an equipment destined for the Flathead trade. Our little party consisted of six "engages" with pack horses, and five armed Indians, amounting in all to thirteen armed men. It was late before we separated from the company, yet, notwithstanding, we marched fifteen miles, killed a fine bull, and halted on the margin of a small spring, in the highlands, near Jackson's Little Hole. We passed through the hole on the succeeding day, and encamped in the narrows below. We fortified our little encampment with a breastwork of logs, or in other words, we enclosed it in a timber pen.

Leaving this, we passed the narrows, a corner of Jackson's Big Hole, crossed Lewis river, ascended the mountains, and on the 30th came into a region where the weather was fair, the sky cloudless above us, and the sun shining pleasantly, quite reverse to the appearance a short distance below. Gazing down, in the direction of Jackson's Hole, from our elevated position one of the most beautiful scenes imaginable, was presented to our view. It seemed quite filled with large bright clouds, resembling immense banks of snow, piled on each other in massy numbers, of the purest white; wreathing their ample folds in various forms and devious convolutions, and mingling in one vast embrace their shadowy substance. - Sublime creations! emblems apt of the first glittering imaginings of human life! like them redolent of happiness, and smiling in the fancied tranquil security of repose; like them, liable to contamination by intercourse with baser things, and like them, dissipated by the blasts of adversity, which sooner or later are sure to arrest and annihilate them. Alike evanescent are the dreamy anticipations of youth, and the aerial collections of vapor. Such the reflections suggested by this lovely scene, which, though often on the mountains, I have never before seen below me. Clouds of this pure snow-white appearance, are, however, by no means uncommon; but those usually observed beneath us, when on the mountains, have a dark and lowering aspect.

Turning with reluctance to things of a more terrestrial nature we pursued our way down to Pierre's Hole, where we fortunately discovered and killed a solitary bull; being the only animal of the kind we had seen since leaving Jackson's little hole.

We passed in our route a well known hot spring, which bursts out from the prairie, on the east side of a valley, near a small willow skirted creek, and flows several hundred yards into Wisdom river. It boils up, at its head, in a quantity sufficient to form a stream of several paces in breadth, and is so hot at its source, that one cannot bear a finger in it for a moment; it gradually becomes cooler as it recedes from the fount and at its lower extremity is cold. The Indians have made a succession of little dams, from the upper end to the river; and one finds baths of every temperature, from boiling hot, to that of the river, which is too cold for bathing, at any season. Our Indians were almost constantly in one or other of these baths during our stay near the springs.

It may be proper to remark here, that we have been drenched with rain, more or less, every day since we left rendezvous. The mornings are generally cloudless and the rocks, mountains, and valleys, are gilded by the dazzling brightness of the sun; but the scene changes as the day advances, dense black clouds cover the face of nature, and heavy rains, though usually of short duration, follow; however, it generally clears up, and becomes warm again, before the sun disappears behind the neighboring mountains. Lightning and thunder are frequent during these storms, and the latter is sometimes distinctly heard, when the sky appears perfectly cloudless.

As we arose on the morning of the 11th of August, we discovered a smoke rising above the pines in the mountains, where we were compelled to pass; however, we packed our horses and started, but saw neither smoke nor other traces of Indians after we had commenced the ascent, we halted at noon on a small prairie, on the summit of the mountain, two hours, and then descended the steep north side, to the head of Bitter Root river. - We saw traces of several horsemen who had recently passed down the river, which is but a small creek here in the mountains, and has a very narrow bottom; we halted late at an old Indian encampment.

Next morning a Flat-head came to our camp, one of several now on the stream a short distance below us, who had separated from a village of Nezperces on Salmon river a few days since, and were now like ourselves in quest of their tribe. Under his guidance we passed down the river into a little valley, then over a high mountain point, and finally descended into a small prairie bottom, where we found the comrades of our guide. We halted a short time to bait our horses, and departed accompanied by the Indians. At dark we halted on a small fork near the river, which has greatly increased in magnitude, and here skirts an open plain, several miles in extent on the east side, but a narrow irregular one, covered with dense pine timber, forming the base of the mountain, on the other. On the 13th, we continued down this river, till evening and halted on it. The Indians with us, announced our arrival in this country by firing the prairies. The flames ran over the neighboring hills with great violence, sweeping all before them, above the surface of the ground except the rocks, and filling the air with clouds of smoke.

On the 14th we met several Flat-heads who, having been informed of our coming by an express despatched two days since, came to guide us to their village, which was situated half a day's journey below. We continued down through an open prairie, and halted late, on the margin of a finely timbered fork, on the west side, a short distance from the river.

On the 15th we passed about seven miles down to an open plain, some twelve miles in length and six in breadth, at the forks, formed by the junction of the Arrow Stone, which flows from the deer-house plains, and the stream we have followed for the last four days; these rivers are both clear, deep, rapid, and not fordable at high water. Bitter Root river is about fifty yards wide, and Arrow Stone something more. The former is but partially bordered by timber some distance above the forks, and the other quite bare; however, in the vicinity of the forks, both branches and the river below, are well decorated by large scattering pines and thick under brush, particularly on the west side. We found the village consisting of only fifteen lodges, situated on the margin of the Bitter Root, about one mile above the forks. As we approached they sallied out from their lodges and shook hands with us, as usual, made short prayers for us during this ceremony, and then passed away as if fearful of troubling us, whilst unpacking and arranging our baggage and lodgings. At length after we had finished pitching our lodge, and disposing properly our luggage, a multitude of women and girls came to us, bearing baskets of fruit and provisions, which they poured upon a blanket and disappeared to their respective habitations. We had suffered much with hunger for several days past, and seated around the blanket, gladly availed ourselves of present abundance to

"Satisfy as well we might,
The keen demands of appetite."

The lovely little hillock, composed of Whortle, Service, Hawthorn, and White berries, rapidly disappeared before the united efforts of eight hearty and hitherto half starved lads; and leaving in it an "awful gap," we arose, either satisfied, or ashamed to be seen devouring so voraciously the gifts of our generous friends; and prepare to receive the Chief and his retinue, who came as customary to smoke a pipe with us, and enquire our business. We seated ourselves in a ring on the grass, with our guests, and a pipe was immediately produced, and presented to a little hardy old veteran, the Chief. He placed it in his mouth, when his attendant applied to it a coal, and the Chief taking two or three whiffs, passed it to the person on his right, who in turn took a few puffs, and returned it to the Chief; he again inhaled through it a few inspirations, and passed to the one on his left, and it continued then regularly round, until it was extinguished. One of the company prepared with tobacco and weed, cut and mixed in proper proportions, and a stick for cleaning the ashes out of the pipe, replenished it, and the same ceremony was repeated. In the mean time we informed them as well as we could, having no interpreter, of our objects in coming among them. They listened without speaking until we had done, when the old chief in the name of his people, tendered us his thanks for coming so far from our chief, through a dangerous country, to bring them munitions and tobacco, articles of which they were much in want; and promised to exert his influence with his young men, to encourage them to hunt and trade with us; supply us with fruit and provisions, and aid us otherwise as much as lay in his power.

He informed us that thirty of his people were massacred last spring, at one time, by a large party of Black-feet, on the east fork of Salmon river. The little devoted band had started expressly to retake horses from, or fight the Black-feet, who were, it appears, approaching in considerable numbers, at the same time, determined to fulfil a threat they had made last fall, that they would exterminate the Flat-heads, root and branch. The two parties met on the summit of the pass from that fork to Horse prairie, and a most desperate conflict ensued, which resulted in the total defeat of the Flatheads, who fought to the last, and perished to a man. The only individual of the party who escaped was separated from the rest in the early part of the action, and fled to tell the disastrous tale. There were among the slain several of the bravest warriors of the nation, who were well known to the hunters as hardy, bold, and heroic in war; sage and experienced in council, and hospitably courteous, even to inconvenience and self-privation, in their humble dwellings. The venerable chief dwelt on the virtues of each of these braves, and related many interesting anecdotes of daring feats and rare presence of mind, exhibited by some of them, in the most trying and hopeless situations. He betrayed the presence of hereditary superstition, in but a single instance: that of a warrior, who had been often wounded in battle by balls, not one of which had ever entered his body, (all having fortunately been nearly spent.) The old veteran declared it the firm belief of himself and friends, that he could not have been killed with a bullet, but must have been caught and then butchered by his enemies. It is believed that the Black-feet sustained great loss in this engagement; at least they abandoned the design of attacking the village, and returned to their own country.

The chief also informed us that he was now awaiting the arrival of the Pend'orielles from the Flat-head post, who were expected in twenty days, and should then proceed to hunt buffalo. These animals are never found west of the Big Hole, or the Deer house plains; consequently the Indians, during their stay here, subsist on dried meat, which they transport in bales of forty pounds or more, upon horses, from the buffalo range, or the haunts of deer and big horn, which are found on the neighboring mountains. They bring in daily horse loads of berries, and several kinds of roots, the most abundant and most prized of which, is small, white, extremely bitter of taste, and called by the Indians "Spathem;" it is found in great plenty in the plain of, and gives name to, Bitter Root river. It is prepared for food by boiling until it becomes like jelly; and is very disagreeable to the palate of one who has never before eaten it. During our stay here, we were employed in trading our goods for the only articles we wanted, namely, beaver and provisions; the best articles in our equipment for this purpose, was cut glass beads, with which the dress of the females are decorated.

Chapter XLII

On the 30th of October we passed under the north east side of the sand mountains which rears its gigantic form in the centre of a vast plain, nearly one hundred miles in diameter, of the most sterile description; where, in gloomy desolation, it seems to mock the vegetating efforts of nature. With the sand hills that constitute its base, the mountain covers an area of some thirty miles in circumference, and is composed of fine white sand, extremely light, into which our horses sank above the knees, in passing over it. When the winds blow upon it, agitating the sand at the surface, it has a beautiful, undulating, wave-like appearance. The particles are so very small, and have so little apparent specific gravity, that large quantities are often removed by the gentle whirlwinds that sometimes occur, and scattered over the neighboring prairie.

On the main body of the mountain there are a few occasional dwarf shrubs, but the hills are absolutely destitute of herbage. No shells, nor any other marine organic remains, nor indeed any foreign substances, are found on the hills or mountains, among the pure chrystals that constitute this singular structure, which, in solitary grandeur appears, when viewed from a distance, like a huge mountain of snow, at any season of the year. Though we have traversed many barren prairies, and crossed over many sterile tracts of territory, we have never before seen any sand of similar character or appearance, which therefore remains unique, and this singular protuberance of earth, an isolated enigma; for conjecture ever must be silent in respect to its origin. It stands alone, in a vast region of elevations, not one of which affords a parallel, or at all resembles it. Is it a proof that the ocean, once slumbering here, deposited this formation, and covered this country, the most elevated portion of North America, situated, as it is, at the sources of the Mississippi, Columbia, and Colorado? If so, the whole region affords no other instance, and even this lacks the strongest evidence of a former subaqueous condition, since there are no productions, or even relics, of an aquatic animal nature attending it. The whole vast plain in which it "towers alone," is covered with the evidences of volcanic action; can it then, in one of those tremendous efforts of nature, which scattered these fragments of rock over so great an extent, have been upheaved from the bed of the earth, and left as a monument of some mighty convulsion? The subject affords ample scope for curious speculation.

Having travelled about twenty-five miles since we set out in the morning, we halted on Henry's fork, about twenty miles above the Forks. Our horses were very much fatigued, and one of them entirely exhausted, was left in the way, being totally unable to proceed, though relieved from the weight of his lading. We discovered the traces, both of a horse and a mule, that had passed quite recently up and down the margin of this river. - The distance from Pierre's fork, when we left the village, to Henry's fork is about seventy miles.

October 31st we remained in camp to rest our weary horses. My companion Newell, however, with one more, departed to ascertain if Dripps had arrived at the Forks. During the day a fire broke out in camp, and filled the air with smoke, but caused no damage.

On the 1st of November, we raised camp, and crossed Henry's fork, but had not left the ford before Newell reappeared with information that a party of trappers, detached from Dripp's company, were now awaiting his arrival at the forks. We continued down the stream about eighteen miles, and encamped with them. Three of them, during a recent excursion for beaver on the sources of Henry's Fork, met with an adventure, which may not be deemed uninteresting, and will also exhibit the danger and hardships to which the trappers are constantly exposed, in this savage region. It is thus described by one of them, William Peterson, of St. Louis.

"On the 23d of September, having returned from my traps unusually early, I set about preparing some meat for breakfast; in the mean time one of my comrades, Chevalia, alarmed or surprised at the continuing absence of Piero, our companion, who had started to go a short distance below, to examine some traps and had not returned; went off to seek him, but in a few moments came running back, out of breath, exclaiming 'Indians! Indians! Indians!' I had barely time to inquire if they were near, when they appeared rushing toward us. We sprang into the willows, where we had taken the precaution to place our baggage when we came here, and I employed myself in arranging it, with alacrity, so as to shelter us from the shower of balls we expected momentarily. Chevalia amused them for some time by talking to them in the Sioux language, which they pretended to understand; happily I discovered at this critical instant, that they were actually taking advantage of the conference, to surround us, and thus cut off our retreat to the mountains; the sudden conviction of this truth, and the fatal consequences to us if they succeeded in effecting their object, induced us to spring at once into the creek, and follow the shallow current about two hundred yards, concealed from our enemies by the few small willows on its margin; when we were compelled to leave it, owing to the total want of obscurity, offered by its woodless borders; and cross a narrow neck of prairie, to a grove of aspen trees on the mountain side. Fortunately for us, the Indians had mistaken the course of our flight, and were some distance below, when we entered the open plain; but they quickly discovered us, and gave chase, as we penetrated the forest; finding that the timber continued we pushed forward, and at length crept into a dense thicket, hoping to elude them thus, but we were again discovered by the barking of a little dog with us, and forced to seek another place to hide in; we finally halted in a dark shadowed thicket, and choaked the dog to silence; while the Indians were yelling like so many deals, and hunting from one grove to another in every direction. We, however, remained quiet and undiscovered, until the sun was setting, when, all the noise having subsided, we ventured out to ascertain if possible, the fate of Piero, who imprudently left us in the morning, without his gun, which was now in my possession.

We returned to our camp, and found every thing removed, from which we concluded that the Indians had departed, and were encouraged to look for the body of Piero, not doubting that he was killed. On our way down to the "dams," (beaver) where we knew his traps had been placed, we passed through a small prairie surrounded by willows; before leaving it, two Indians intently gazing on the ground, were observed, evidently following our tracks. I immediately placed one of my guns between my knees, and shot the foremost of them with the other, dead on the spot. My companion fled, and the surviving Indian slowly brought his gun to his eye, as if to insure my fate; but in a twinkling, the empty gun dropped from my hand, and the other was at my eye. The amazed red skin sprang into a thicket and escaped. - At this moment a volley of diabolical yells burst upon my ears from the adjacent willows. I seized my empty gun and ran for life, expecting to experience something more fatal, though scarcely more appalling than the horrid sounds that saluted me at every step. Unharmed, however, I reached our former place of concealment, to the no small gratification of my companion, whom I believe, had hardly deigned to bestow on the Indians a single glance, from the moment I fired until he found himself safely and snugly in our favorite quarters. During my flight I saw many of them, both in the little plain where I shot one, and in pursuit of me, but I could form no correct idea of their sum total.

After dark all became quiet, and we sallied forth from our hiding place, passed over the plain four miles to a neighboring stream, and remained there until morning, when we returned to the scene of our misfortune, with a slight hope that Piero had escaped the knives of the savages, and might be awaiting our reappearance. But as we approached camp they were again seen, and we fled to the mountain unperceived by them, giving up all expectation of ever beholding our comrade. The Indians were encamped a short distance below, on the margin of the stream, and we were in constant danger of being discovered and butchered by them. At length, however, we set out for snake river, and reached it after two day's march, without accident; but being unprovided with horses, and meeting no game, we were four days without food. Good fortune on the fifth, put a fat cow in our possession, and we remained to recruit ourselves until the following day, when we set out for the Buttes at the forks of snake river, to find, if possible, a party who were trapping that stream in canoes. After an anxious search we could perceive no traces of them, and concluded to return back to the cow we killed the day previous, and wait for the arrival of the company at the forks, as patiently as possible.

During our progress we discovered, with surprise, the tracks of a white man, a close examination of which, showed that his feet were widely different from each other. It occurred to me at once, that Piero had a deformed foot which I recollected to resemble the print of the unnatural one here. The hope of finding the old man yet alive, burst forth anew, and impelled us forward with fresh vigor in the direction indicated by his footsteps, which we followed a distance of two miles, that seemed but a step, filled as we were with the pleasing anticipation of shortly relieving his miseries. With a joyful swelling of the heart, that it is absolutely impossible to express, and that can only be imagined by such as have felt the thrilling ecstacy inspired by circumstances similar, I perceived my old friend and companion, on the margin of the river, at the distance of about half a mile, and fired off my gun, the usual token of delightful recognition, to attract his attention; he turned toward us - extended his arms in a transport of glad surprise - and overpowered by exhaustion and the excitement of feelings it were impossible to control, sank fainting on the ground. We ran to him - raised him up - chafed his temples; he recovered - clasped us to his heart - tears flowed profusely down his aged cheeks, and utterance entirely failed him. Never had I before, never can I hope again, to experience such another moment of melting tenderness and joy. I could, I think, face death with heart and courage unappalled, not a drop would moisten my eye, and not a fiber tremble, but the scene I have vainly attempted to describe quite unmanned me, and our tears were mingled together. We supported him with us to our camp, tended him with affectionate care; for if there is in this world any circumstance, that can soften the heart, and awaken our affections to each other, it is a companionship in misfortune; but we could not deny him the gratification of eating until he had so much overcharged his stomach, as to make himself quite sick, from which he did not entirely recover in seven or eight days; during which time we killed seven buffalo, and fortunately fell in with a party of trappers, with whom we have since remained."

Chapter XLIII

From Piero, whose real name is Jacques Fournoise, the former appellation having been given him by the Canadians, most of whom have nicknames, I received an account of his sufferings, which will, I hope, be a sufficient caution, to strangers particularly, who may hereafter visit the Rocky Mountains, never to leave, without their arms, their encampment or companions.

It appears that, on the morning before alluded to, he started for his traps with two beavers, both very large, on his way back to camp; and observed as he was about leaving the willows, a number of buffalo robes scattered about on the prairie. The startling Ruth flashed on his mind, like an electric shock; he regretted deeply his imprudence in venturing from camp without his gun, which left him now without the means of defence, or subsistence, and dared not attempt to reach it, where, in all probability, his comrades already were shipped, scalped, and covered with wounds - dead. He heard the sanguinary yells of the savages re-echoed back from the neighboring hills and mountains, and was induced, by his own peril, to seek concealment, which he finally effected in a dense cluster of willows. He heard the footsteps of his foes passing and repassing frequently during the day, but in the evening all was silent and he left his shelter to revisit camp. Arriving there, comrades, horses, baggage, all had disappeared, and he remained alone, unarmed, a wanderer in the barren plains of Henry's Fork, with misery, starvation, death, and such a death pictured to his imagination in their most gloomy colors. With such a vivid prospect of despair, he carried his beaver some distance up the stream, split the meat for drying, and remained in cheerless solitude during the night.

On the following morning early, he revisited his traps and found in them two beaver, with which and two of his traps, he returned; by the aid of the latter he hoped so to increase his store of provisions, as to render more remote the danger of perishing with hunger, and enable him, if possible, to await the arrival of the company at the Forks, where they were expected to be by the first of November. The reason that he did not meet with the Indians upon the night and morning that he visited the place of encampment was because he did not, in the first instance, leave his place of concealment until it was quite dark, and in the second, he went and returned before they were stirring.

He moved into the mountains several miles, when he dried his meat, and then sallied forth again into the plain, crossed Henry's fork, and proceeded with the intention of hunting beaver on a small branch. At length he found a dam, and leaving his meat, traps, etc., went across to examine it, but reached the opposite side just in time to discover himself to two parties of the copper-faced rascals, who were situated both above and below him, and who sprang up in chase as soon as they saw him. His fate for a moment seemed inevitable, but the energy of despair lent vigor to his aged frame, and activity to his feeble limbs, and with an agility that would have done credit to his youth, he ran across the bottom, several acres in extent, only covered with here and there a few scattering willows, in the hope of finding a dense thicket; when he reached the border of the plain, and heard the footsteps of the Indians fast approaching, death appeared inevitable. The child of luxury would unquestionably have sank under the weight of opposing circumstances, and making no further effort to prolong a life equally dear, have fallen an easy victim to savage atrocity; but the hard buffetings of a mountain life had familiarized the old veteran to danger, and impressed it on his mind as a duty, never to yield to adversity, while a shadow of hope remained; and to scan, even in the most critical moment, the chances of success: his experience was even here not unavailing; he sprang into a cluster of willows, and crouched with deliberate caution, at the very moment the Indians passed him, pulling and tearing the branches asunder in every direction. Some of them at the same time, commenced making forts or lodges but a few feet from the spot where he was concealed; and at every moment he expected part of the branches that surrounded him, would be cut away for that object, and thereby expose him. It was still morning, and he studied to calm his agitation, and meet a doom he had reason to think unavoidable, with manly fortitude.

In the meantime the search continued with such unabated ardor, that he believed every bunch of bushes on the whole bottom, underwent an examination, except in the single instance that afforded him protection. Towards noon some of them commenced playing roulette, close under him, against the willows; others seated themselves around, and in one instance he felt the body of an Indian touching his own. His heart beat so violently that he fancied they must hear it, and would soon send him to eternity. Shortly one of them sprang up, seized a tin vessel, struck it thrice on the bottom, and started towards him. He now thought the Indian had discovered him while sitting by the bushes; but wished to make it appear to his comrades, as if magically effected, by virtue of the kettle. At this trying moment, his soul rose superior to his condition, and he had the presence of mind to consider that they might torture him if taken alive, and the firmness to prefer present death to prospective hours of agony; he grasped his knife, and determined to spring upon the first who approached him.

"Poor powerless man, and canst thou harbor then
A hope but death - a passion but despair."

The Indian, however, passed him, examined several other clusters near, but shortly returned and again seated himself near his former position, and commenced crying, in the manner of those people, at the death of a friend. The day passed slowly away, the hours to him were attenuated to many times their usual length, and he almost imagined night would never approach.

"In those suspended pangs he lay,
Oh! many, many hours of day!
How heavily it rolled away."

And slowly night advanced; but it came at last, and the bloodhounds retired to slumber. A spark of hope rekindled in his bosom a new energy. He carefully removed the branches to make his escape from that tedium of terror, but as he was about to step forth from the thicket, the Chief hurried out of the fort, and harangued his people for some time; at length he reentered his lodge, and in a short time all became quiet. Our hero seized the opportunity, cut the willows until he was able to extricate himself without noise, and then carefully crawled out into the prairie; his legs had been cramped up all day and now refused to perform their usual office. With extreme difficulty he succeeded in dragging himself on his hands and knees some distance and after rubbing his limbs considerable time, to restore and accelerate the circulation of the vital fluids, he was enabled to resume an erect position, and walk off as fast as his years would permit.

During this painful midnight march, a large grizzly bear reared up on his hinder legs but few feet distant. - At any other time he says he should have been alarmed to meet so dangerous an enemy, but his feelings had been so lacerated by constant peril, that no new danger could appall him. With more of forbearance or pity, than his human foes had exhibited, the bear turned and retired, without adding brutal molestation to man-created misery; and Piero quietly pursued his way to the forks of Snake River, where he wandered about for some time in search of roots and buds to support his existence.

"Poor child of danger, nursling of the storm!
Sad are the woes that wreck thine aged form."

Fearing at last that the company would not come to Snake River, he resolved to retrace his steps, and endeavor to find some of his traps. During the chill frosty nights, having no bedding, he cut up and covered himself with grass, his only means of shelter.

Finally, on his way back he was discovered, as we have before related, having eaten nothing for six days, save a very few roots and buds. His companions, judging from his emaciated condition when found, think that a few more days would have sufficed to close his earthly career. So wretched and hopeless was he, that when he heard the joy inspiring report of a rifle, and turned and saw friends, comrades, approaching him, the revulsion of feeling from despair to the fulness of hope, was too much for feeble nature to sustain, and with a single gesture of gratitude, and one wild loud throb of exultation, he fainted. With the same sensation, described by that master spirit, Byron, he must have thought, when he came to, and found himself in the arms of his companions -

"Awake! - where am I? Do I see
A human face look kind on me?"

Yes, his sorrows were past, his woes forgotten, and for once he gave himself up to the luxury of gratitude and joy. The soul of the mountaineer was subdued, the warm tears coursed each other down his aged cheek; his limbs trembled with the vibration of ecstacy, he could not stand, and they all sat and wept and laughed together.

Chapter XLIV

There were at the Forks three other small parties, who had been compelled to leave their traps and fly for safety, from the Indians. However they have since returned and collected them all, save forty, that probably fell into the hands of the savages. On the 2d of November Mr. Dripps arrived with the company, having been quite fortunate during the fall, and caught many beaver, losing neither men nor horses. He had seen no Indians, with the exception of a party of Snakes, and had received information from some source, not recollected, that a detachment of Bonnyville's company were attacked by the Crows on the head of the Yellow Stone river, and lost several horses, laden with merchandize, having also two men badly wounded. Likewise that Mr. Cerre, a gentleman of that party, had departed down the Big-horn in bull-hide boats, with the peltries thus far collected by them, destined for St. Louis. We heard, in addition, that the Crows had discovered a cache belonging to the same company, and robbed it; however, they must have been seen while constructing it, or it could not have been properly closed; for a well made cache is in no more danger of being discovered in these plains, than is any substance surrounded by the fibrous effects of organic nature. We were also told that Mr. Thomas Fitzpatrick at the head of a detachment of the R. M. F. Co., had their horses nearly all stolen by the Crows; which, however, they afterwards recovered; and report adds, that two of his men were killed. Capt. Walker was seen on Snake River, some distance below. His men had lost some traps taken by the Rootdiggers, and had killed one of them in return. The Crows have openly declared war against all hunting parties found trespassing on their territory. A party of trappers, detached from the R.M.F. Co., are also on Snake river; one of them, named Small, was killed by the accidental discharge of a gun some time since.

Trading houses have been established at the mouth of the Maria, and also at the mouth of the Big-horn by the American Fur Company; who likewise intend to erect one at the three forks of the Missouri, for trading with the Blood Indians. From the Indian signs we saw at the deer house, when with the Flat Heads, I concluded that this fort was already constructed - judging by the bits of blankets, cloth, etc., and other emblems of commerce observed there. From Mr. Fontenelle, and some others, no intelligence has been received since they left rendezvous.

On the second day after the arrival of Mr. Dripps, we selected a good encampment, affording wood, water, and grass in abundance, situated on Pierre's Fork, near the mouth, and removed to it. In a few days Mr. Pillet prepared to set off on his return to the Flat Heads; and having received some encouragement, I concluded to go with him: accordingly we started on the 12th; our little party was composed of some seven or eight white men and five or six Indians. Not a few of our friends, deeming our journey, in such small numbers, extremely perilous, expressed an opinion that we would be massacred by the numerous parties of Blackfeet, who ever haunt the country, through which we would have to pass, at this season of the year. However we were not dismayed, and marched over the hills, mountains, and plains with all possible dispatch, i.e., as fast as our fatigued and feeble animals were able to proceed, and reached the Flat Heads on the evening of the 27th, at Bitter Root river Forks. We followed the same route Newell and myself had pursued the previous summer, and which is indeed the nearest and best. With the single exception of cold winds, we were favored with pleasant good weather during our journey. But the nights were exceedingly cold, and our blankets and robes absolutely necessary. We found buffalo numerous the whole distance from Dripp's camp to the Big Hole, and we therefore lived well.

On the north side of the mountain we crossed, there was already one foot of snow, but the plains were yet entirely free. We saw several encampments made by the Flat Heads, since I left them; in one of these we saw a living colt with one thigh broken by a ball. At the mouth of the defile, (from the Big Hole to the Bitter Root,) we found several hundred lodge poles, left by them before attempting to pass over the mountains. One of them, painted red and black, was planted erect in the ground, and was probably intended to be emblematic of defiance to the Blackfeet. We caught on the mountain a fine horse, that had been lost by the Indians, and took him on with us.

On the east side of Bitter Root river, there is a singular curiosity, that I had not before observed, because it is situated under some rocky bluffs, almost impassable to horsemen, the proper road being on the west side of the river: it is the horn of an animal, called by hunters, the "Big-horn," but denominated by naturalists "Rocky Mountain Sheep;" of a very large size, of which two-thirds of its length from the upper end, is entombed in the body of a pine tree, so perfectly solid and firmly, that a heavy blow of an axe did not start it from its place. - The tree is unusually large and flourishing, and the horn in it some seven feet above the ground. It appears to be very ancient, and is gradually decomposing on the outside, which has assumed a reddish cast. The date of its existence has been lost in the lapse of ages, and even tradition is silent as to the origin of its remarkable situation. The oldest of Indians can give no other account of it, than that it was there precisely as at present, before their father's great grandfathers were born. They seldom pass it without leaving some trifling offering, as beads, shells, or other ornaments - tokens of their superstitious veneration for it. As high as they can reach, the bark of the tree is decorated with their trifles.

We met with a few Indians before reaching the Forks, who were hunting deer a short distance below the Sheep horn; and some of them accompanied us to the village, when we found Mr. Ermatinger and most of the Flat heads. His men, with the Pen-d'orilles, had departed towards the Flat-head post, and he had remained behind, for the purpose of seeing us, having been informed of our approach, by an express despatched by Pillet, a day in advance of us. From him we learned that a young man had been killed near the Big Hole, since we left the villages, by the Blackfeet, and that the Indians had lost several little bands of horses; most of the young warriors being now in pursuit of the robbers.

On the 29th we again set out, in company with that gentleman, and passed northward through this valley and a narrow defile that led us out into a very small plain on the mountain, or rather in it, called Little Cammas Prairie, having made about fifteen miles. Mr. Ermatinger and myself departed on the succeeding morning, to overtake his men with the Pen-d'orilles. We descended the mountain into a fine valley called Cammas Prairie, watered by a beautiful, well timbered stream flowing northward; passed through it, and thence along a narrow defile where the river is confined on both sides by hills rising, though not perpendicularly, from the water's edge many hundred feet, and at length overtook the company who were in the act of encamping when we came up, having rode not less than twenty miles. The same evening Pillet arrived, and halted with us, and about the same hour an express came in from horse plains, with information to Mr. E. that some of his trappers had arrived there. Shortly after dark two Indians reached us from the Flat-head village, bearing a scalp, and informed us that the party who went in pursuit of the horse thieves, had returned with twenty-six of the stolen horses, and two trophies from the heads of their enemies.

Chapter XLV

December first, we proceeded again on our journey, and found ourselves after a march of twelve miles, on the margin of a large clear river, ornamented with scattered pines on its borders, called the Flatt-head river.
On the 2d we succeeded, after considerable difficulty, in fording it, still deep and rapid, even at its lowest stage, and left it, passing through a large prairie, and finally halted at the entrance of a mountain pass, having marched about fifteen miles.
On the succeeding day we continued our toilsome progress over a low pine covered mountain, from the summit of which we descended into a large river valley, called Horse plains, watered by the Flatt-head river, which here joins with the Bitterroot. On the margin of the former, we found a party of young half-breeds, who had returned from their fall hunt, and brought in their furs for delivery to Mr. Ermatinger. These men are supplied with goods on this, the west side of the mountains, by the traders here, much cheaper than the Americans can afford them on our side; and consequently they are better clad, though, for want of buffalo, they do not fare as well as our hunters. There was likewise quite a village of Indians collected here, who never go for buffalo, nor do they in many cases possess horses to go with, even if they felt inclined to do so. Most of the families, have light canoes, with which they glide about on the river, and gather roots and berries, in their proper season; but in the winter they separate into small parties, and not unfrequently, into single families, who then seek the mountains, and pass that inclement season there, with little knowledge of or communication with each other. - They assemble here, at certain seasons to

exchange with the traders their skins for such articles as they may chance to want, or be rich enough to buy.

There was a cabin once erected in this valley, for the purpose of trade, but it was shortly afterwards consumed by fire. These Indians were anxious to trade for dried buffalo meat, and I think in many instances succeeded in exchanging some of their fish and roots, for more substantial viands. I had the pleasure of tasting some vegetables that had been conveyed here in two large open boats or barges, with goods from an establishment seven or eight days' journey below, called "Colville," at the Kettle falls.

When we arrived in this valley the danger was declared over, and our horses permitted to run loose night and day; however, a few nights after, thirty of them were stolen. Mr. Ermatinger lost eight or ten, and the others were the property of the Indians. Mine, as good fortune would have it, were not among the number missing.

On the 13th of December Mr. Ermatinger, having completed his arrangements, and stowed his furs into the barges, which were manned by his men, who were, I perceived, quite "at home" in them, set out with his companion, Pillet, for Colville. In the mean time a general breaking up of camp took place. In a few moments the lodges disappeared, and the bosom of the river was studded with bark canoes conveying whole families and their baggage down the stream with surprising velocity. I was greatly deceived in their canoes, for the squaws would lift them from the water on to the bank, and again set them into it, with such ease that I imagined they must be quite insufficient to the transportation of any heavy burden. Some of them, however, appeared loaded until there was no longer room for any thing more, and still floated securely. They were managed by the squaws, who, with paddles, direct their course with great steadiness, astonishing rapidity, and apparent ease and dexterity. Parties on horseback, at the same time set off in various directions, and finally I departed, with the family of a trader, to pass the winter in the Cotenas mountains. We travelled this day over a low spur, or point of mountain, east of Horse Plains, and halted in the pines on the margin of a small stream, that flows into the plains.

After a day or two, we passed into a small, though open valley, watered by a stream flowing westward, called Thompson's river, in honor of a gentleman formerly in the Hudson's Bay Co.'s service. We had several days fine weather, when Heaven seemed to be literally "smiling over us," and much enjoyed it.

At length, on the 20th, two "Canadian Voyagers" came to us from the Flatt-head post, where they had been in quest of some articles of commerce. They remained with us a part of the day, and set out for their place of destination. Two Pen-d'orielles Indians came and engaged to hunt for us; but one of them left us several days after, promising to return again soon.

In the mean time Mr. Montour and myself, set about constructing a log cabin, rather because we had nothing else to do, than from any necessity, as his lodge was uncommonly large, and quite comfortable. Christmas was passed agreeably with the family of mine host, and we were rather more sumptuously entertained than on ordinary occasions. Our "bill of fare" consisted of buffalo tongues, dry

buffalo meat, fresh venison, wheat flour cakes, buffalo marrow, (for butter,) sugar, coffee, and rum, with which we drank a variety of appropriate toasts, suited to the occasion, and our enlarged and elevated sentiments, respecting universal benevolence and prosperity, while our hearts were warmed, our prejudices banished, and our affections refined, by the enlivening contents of the flowing bowl. Our bosoms glowed with the kindling emotions, peculiar to the occasion. - Remote from our kind, and the thralling, contracted opinions which communication with a cheating world are apt to engender, when our stomachs were filled with substantial viands and our souls with contentment, we were at peace with all mankind and with ourselves, and had both time and opportunity to expatiate largely on honesty, charity, and philanthropy; which we did till our goblets (tin cups) were drained of their inspiring contents, and night summoned us to repose. Do not imagine, reader, that as we slept off the effects of our conviviality, we also slumbered away these enlarged and liberal views of ethics, honor, and integrity. No! With a praise worthy propriety, they continued to increase and multiply, until the next favorable opportunity offered for taking advantage of the ignorance and necessity of the Indians, in honorable barter; when, having occasion for them, they could no where be found, but had vanished, and like "the baseless fabric of a vision, left not a wreck behind." Sublime and intellectual as were these enlightened principles of morality, how vastly to be deplored is the fact that they are not more generally and permanently entertained. Alas! for poor human nature! Truth is too abstract and difficult to be comprehended - Justice too holy and intricate to practice - Honor too lofty and profound to be governed by - and all too obsolete and unfashionable to direct - in the vulgar concerns of trade. The most upright principles, the most equitable regulations, the most honorable expedients, all, all retire from their formidable enemy - selfishness; and intrigue is but another name for frailty, when opposed to the potential influence of interest. Such, at least, I have found the world though there are doubtless many, and I have met with some most honorable exceptions.
"*Auro pulsa fides, auro venalia jura.*"

Chapter XLVI

On the last of December, our remaining Indian hunters left us, promising to come back again in a few days, however, they never returned, - and the day passed as usual.
So ends the year 1833, but let it pass in silence to oblivion, with the thousands that have gone before, and must hereafter follow it. Its hopes, its fears, its expectations, can no longer excite or agitate; its perils, and privations, are already half forgotten; and even the severe disappointments it produced, have entirely lost the ability to disturb. Well! with all its imperfections on its head, let it descend to the dark void of by-gone eternity, and we, older if not wiser, will

again look forward with eager curiosity, to the events which yet slumber in the bosom of another; with wishes that will probably prove to be vain, with hopes that may be destined to experience disappointment, and with expectations that are doubtless doomed to be blasted; but which are all, notwithstanding their frailty and uncertainty, cherished with fondness, perhaps with folly.

The new year was ushered in with feasting and merriment, on dried buffalo meat, and venison, cakes and coffee; which might appear to people constantly accustomed to better fare, rather meagre variety for a dinner, not to say a feast. But to us who have constantly in mind the absolute impossibility of procuring better, and the no less positive certainty, that we are often compelled to be satisfied with worse, - the repast was both agreeable and excellent; for think not, that we enjoy, daily, the same luscious luxuries of cake and coffee, that announces the advent of 1834; by no means. Our common meals consist of a piece of boiled venison, with a single addition of a piece of fat from the shoulder of the buffalo; except on Sundays, when we have in addition a kind of French dumpling, made of minced meat, rolled into little balls enveloped with dough, and fried in the marrow of buffalo; which is both rich and pleasant to the taste.

Our house was now advanced to putting on the roof, for which we had cut a sufficient number of poles, intending, when properly placed, to cover them with grass, and finally with a coat of earth, sufficient to exclude rain or melting snow, in the spring.

On the 13th of January snow commenced falling on the hitherto uncovered earth, and continued without intermission for five days and nights; when the storm suddenly abated; and we enjoyed fair sunny weather, every day, during the remainder of the month.

In the mean time we had finished and moved into our house, which was rendered extremely warm and comfortable, by having the seams filled with clay, a chimney composed of sticks and mud, windows covered with thin transparent, undressed skins, which admitted sufficient light, and yet excluded the rain and snow; and a floor constructed of hewn slabs. After the building was completed, and the whole family snugly housed in it, the son of Mr. Montour, a young man of about nineteen, and myself employed ourselves in hunting deer on snow shoes; but in consequence of our not having rifles, our sport was quite limited.

We each possessed a fusil brought to this country expressly for the Indian trade, a light kind of gun which is used only by the hunters on our side of the mountains for running buffalo. However, my companion appeared quite expert with his weapon, and made several very good shots with it.

During these rambles we sometimes saw an animal resembling an otter, in size, shape and color, called a pekan or fisher; but for want of a dog to tree them, did not shoot any. We killed several martins, and saw the traces of a large animal of the cat kind, supposed to be a lynx.

At length, my companion and myself, having become tired of hunting, concluded to go to the Flatt-head house, where we knew several families of Indians were wintering, and engage one or two, to come and hunt for us. - We started on the last day of January, passed down Thompson's river about ten miles, and encamped on its margin, in good time to clear away the snow several places in extent, prepare for a bed the branches of the balsam-fir, which we cut from the trees that abound here; and last, though not least, to kindle a fire and heap on the wood in plenty, which we did, and made, what would be called at home, a rousing one. In the evening we dried our moccasins, ate some dried meat, and finally threw ourselves down to sleep, covered by our two blankets; and as the night was pleasant we rested well, warm and comfortable.

Quite refreshed, we rose early next morning, and continued our journey. We went down the river several miles farther, killed a deer and encamped near it, in consequence of a severe storm of rain which commenced in the forenoon and continued all day. We however went out, killed another deer, and trailed it along the snow, with a great deal of difficulty and fatigue, to camp. After reaching it, we constructed a shelter of fir branches, aware that it would not exclude the rain, but in hopes that it would change to snowing. Placing again, a quantity of branches on the ground, we lay down upon them "to sleep, perchance to dream," both of which paid us visits, but neither of them remained long.

Next morning, February 2d, we rose, as the "Velchman" said to his "Vife," "vell vet," being literally drenched to the skin. The rain was still falling very fast but the air was warm, and the snow rapidly diminishing. We had some trouble to kindle a fire, every thing being wet. However, with some splinters of pine full of pitch we finally succeeded, and had soon a good fire. By it we dried our clothes, and then set out in quest of game, although it was still raining. I passed down the river some distance, and fired several shots without effect, but my gun soon became so wet that I could no longer get it off, though I had several opportunities, and was obliged to return to camp. My companion had the same ill luck, and was already there. We each saw the bones of several deer that had been caught and devoured by wolves. The wind was blowing from the south, and the rain still falling, when we again lay down for the night; not as people do in some parts of the world, dry, warm, with plenty of bedding, and a shelter from the storm; but wet to the skin, exposed to the pelting rain on a February night, in a latitude of about fifty degrees north; our only covering a couple of blankets, that would scarcely, if immersed in a fountain of water, have robbed it of a drop, and our bed the bosom of our dear mother earth, only overspread with a few wet branches, where we lay, and a fine coat of melting snow where we did not. The little cool rivulets, trickling down every part of us, and gliding along the skin on our backs, prevented sleep from injuring our delicate faculties, and kept us fully awake to the very romantic situation in which we were placed. However, the air was not very cold, and we rose rather early, if I remember right, for this fact was not recorded in the diary I kept while in the mountains, kindled a fire, and partially dried our clothing. We cooked a shoulder of venison and ate it for breakfast, and then, considerably refreshed, set out again in quest of game. I succeeded in killing a young buck, but my gun soon became wet and useless and I set out for camp, dragging the body of my deer along on the snow. On the way my companion joined me, and having been unsuccessful, assisted me with it to camp. In the evening, we came to the conclusion to abandon, for the present, our design of going to the Flatt-head house, because the snow had become so soft that it was almost impossible to get along with snow shoes; and quite so without them.

We resolved to go back, and get horses to convey our present supply of provisions to the house. After dark we again lay down on a bed that could not be mistaken for one of down, and in a plight that will not, I presume, excite envy in any one, being precisely similar to our condition on the preceding evening, as it yet rained steadily, though slowly. Do not flatter yourself, reader, that we slept any, though we had been without for the last forty-eight hours; that was a luxury only contemplated in our extremely interesting situation; our enjoyments were entirely intellectual, and we could not abandon the pleasures of reflection even for a season, to waste our precious time in sleep.

The night, though lengthened to a degree that we would not have conceived possible, at last approached to a close.

We kindled a fire, prepared a scaffold to secure our meat from the wolves, placed a handkerchief as a flag above it, to frighten away the ravens and magpies, which are always numerous in this country, slung our blankets, and started for the house. The snow was now reduced to the depth of a foot, but very soft and saturated with water, so that we sank through almost to the ground at every step, even with our rackets on. However, we reached the house after dark, wet through, and nearly exhausted by fatigue; changed our apparel, and having ate a hearty supper, retired to enjoy a good nights rest in our beds. Oh! the luxury of such repose! after having been exposed to a storm of rain, for three whole weary days, and two sleepless nights. None can appreciate it, who have not experienced similar hardships.

It rained all night, but abated at the dawn of day, when

"Up rose the sun: the mists were curl'd
Back, from the solitary world."

and the mellow rays of the sun again illumined nature with their cheerful splendor.

Chapter XLVII

On the morning of the sixth we mounted two of our horses, and set out for our late encampment, where we had enjoyed so many hours of nocturnal contemplation. - The day was one of the finest imaginable, clear and warm, and every object tinted with the rich lustre of the radiant sun. Late in the afternoon we reached our venison, which remained as we had left it, untouched by beasts or birds. During our progress we had seen several deer, but could not get a shot at them. We kindled a fire and prepared a shoulder of venison for supper; but had no water, and being some distance from the river, could not get any without descending a steep rocky bluff, at the imminent hazard of our lives. We tasked our ingenuity to devise an expedient, and succeeded. Taking the digestive ventricle of a deer, which had previously been cleaned for food, we filled it with snow, and then putting in heated stones, continued the operation until we were at length the lawful proprietors of a whole gallon of water; which auspicious event, was duly celebrated by proper rejoicings.

Having finished these important arrangements, eaten supper, and participated in the refreshing beverage so ingeniously procured; we laid down on the precise spot where we had spent three previous nights, and enjoyed an undisturbed repose.

On the day following we returned to the house, with another deer which we had the good fortune to kill by the way. The weather continued pleasant for some time forward, and on the tenth, two Indian boys came to us from a lake, two or three day's walk to the eastward; where the half-breeds we saw in the Horse Plains, were passing the winter. A child of one of them was burnt to death accidentally, by its clothes, which were of cotton, taking fire. In that quarter there had been as much snow, as in our vicinity. Next day the boys continued on, in quest of some relations at the Flatt-head house. - We told them to send up here one or two Indians to hunt for us, and they should be well rewarded.

We waited until the 17th for the arrival of some, but none came, and I finally resolved to go down and engage one or two myself. Taking my gun, blankets, a pair of extra moccasins, flint and steel, a hatchet, and my snow shoes, I passed down the river about twelve miles, where I found a comfortable bush cabin, and halted for the night.

In the morning the weather was pleasant, and I started down the serpentine course of the stream, hemmed in, on either side, by an indefinite number of high rocky bluffs, or points, with sides almost perpendicular, jutting into the river; on which, I was frequently compelled to climb, at the hazard of my neck. Had I followed a guide, no doubt much fatigue, danger and distance, would have been avoided, but ignorant as I was of the proper route, I was compelled to follow the tortuous course of the river; often to retrace my steps, and seek a more practicable passage, from some abrupt precipice, or perpendicular descent; till at length, quite overcome by fatigue, I sought refuge from the storm - which recommenced, soon after I started in the morning - in a cavern, which extended far beneath an immense rock, and afforded a comfortable shelter. When I halted here it was quite dark, and I was wet and benumbed with cold; however, I gathered a quantity of fuel, and succeeded, after considerable difficulty, owing to the wet condition of every thing I procured, in making a fire, the cheerful blaze of which, illuminated the interior of the cave, and enabled me to discover an excellent spring, a few paces in the interior, precluding thus the necessity of climbing down the steep bank of the river, fifteen or twenty feet to the water's edge, and exposing myself to the risk of falling in, of which there would be great danger. During the day, I had no opportunity to kill any game, nor did I see any signs of deer.

Fasting, and unrefreshed, I set out early on the morning of the 19th, scarcely able to see, in consequence of the snow, which was falling with almost blinding rapidity. I had, however, made but a short distance, when I discovered the traces of two Indians; and proceeding but a little way further, found two squaws in the act of butchering a deer. By signs, they informed me that I would find several Indians a short distance below. - I continued on, till I found about fifteen men and boys, warming themselves by a fire of large dimensions. - They were from the Flatt-head house three days since, and came here to hunt; they directed me to follow a small trail, which they pointed out in the snow, and that I would find a house. I traced along the path about half a mile, and came to a large bush cabin. In the evening the Indians all returned, having had pretty good success in their hunt; and we slept in a circle not more than ten feet in diameter, with a fire in the centre; though there were seventeen souls in all, not estimating in this number two dogs, as their having souls is yet a matter of dispute with metaphysicians. They, however, considered themselves as of us, by insisting on places in the circle; which, notwithstanding sundry kicks, cuffs, and other dogmatical demonstrations of hostility, they retained during the whole night.

In the morning following we started for the house, passed out into the plain, at the mouth of Thompson's river, several miles in extent, and occasionally intersected by woodland. Our party was divided in this place; two of them I despatched to the house of Mr. Montour; and, with the exception of two others who accompanied me, the remainder departed down the river. Under the direction of my guides, I passed across the plain several miles, to the Flatt-head house, situated on the river bearing the same name. This establishment formerly consisted of seven hewn log buildings; but all are now going to decay, except the one inhabited by the Indians who accompanied me. They supply themselves with firewood, at the expense of the other buildings. They are entrusted with the secret of a cache, made in one of those decayed houses, containing the goods which yet remain of Mr. Ermatinger's stock last fall. He would be there shortly, they informed me, on his way to the plains. During my stay at the fort, the Indians went out daily to hunt, and seldom returned unsuccessful.

On the 24th, an Indian by the name of Pillet (so called after the Trader) came and invited me to go and sup at his camp, which he assured me, was not far distant; I gave a ready consent, as he would be greatly obliged by my compliance, and I would not wound the feelings of a Pen-d'orielle by a refusal, if it could conveniently be avoided. We passed down the river some distance, crossed it on the ice, and continued over hills and hollows, through a grove of pines, about four miles, to his lodge; which was situated on the margin of a small stream in the mountain, and was constructed of weeds. Though it appeared better calculated to exclude the warm rays of the sun, than to keep out the cold; a cheerful fire within counterbalanced the evil; and I was seated opposite to a good natured squaw, and two or three children. My considerate host set before me a vessel, filled with the choicest morsels of deer and lynx, both fat and tender, of which I partook freely. The flesh of the latter is far superior to the meat of the deer; and is the best I ever tasted, except that of the female bison. After finishing my repast, I ascertained that he had passed the winter thus far here, with his family, quite remote from any of his people; that he had killed, during his stay here, forty-six deer, two lynxes, two pekans, three martins, one beaver, one otter, and several muskrats. He appeared quite intelligent and devout, prayed for his family before eating; and so pleased me, by his courteous yet dignified conduct, and amiable disposition, that I concluded to remain at his lodge, until the arrival of the trader. Time passed agreeably here, and I made considerable progress in acquiring the Flathead language; in which my kind entertainer seemed delighted to instruct me.

Chapter XLVIII

On the first of April, Pillet departed for the Cotena trading house, in company with the three men from that place; and on the same day, a half-breed arrived, with information, that his friends would be here in a day or two from the Flatthead lake.

On the eleventh an Indian reached us in advance of a party coming on from Bitter-root river; he said that three Flatt heads had been killed in that quarter, by the Black feet. The Indian, Pillet; whom the reader will recollect as my kind entertainer a few days since, came here on the twelfth, to receive a present I had promised him in return for his hospitality. He informed me with sorrow, that "Tloght" had been severely wounded in a contest with a lynx, and that he was much afraid he could never recover. I felt really grieved for him, knowing the value of the friend he was likely to lose; but lest the reader should feel an emotion of regret that he might afterwards think misplaced, I hasten to inform him that Tloght, the poor wounded Tloght, was of the canine species. Yes, reader! he was, though one of the noblest of his kind, only - a dog - a large powerful black hound, exceedingly swift, and well trained to hunting. He has often caught and killed deer of the largest kind, unaided; he was also fond of hunting lynx, and animals, very strong and fierce; and indeed, I have been more than once indebted to the courage and agility of Tloght, for a most delicious meal of the flesh of this animal. On one occasion, when Pillet was hunting with his faithful dog; the latter treed a lynx - which the Indian killed, - and there found in the snow beneath the tree, no less than three fine deer, which the lynx had killed, and buried there; all of which, through the sagacity of his dog, became a prize to the hunter. The name - Tloght - which was borne by this fine animal, signifies Fleetest; and is but a literal expression of his speed. Pillet, was a poor Pend'orielle, whose sole reliance for the support of his family and himself, was on his dog, and his own exertions. He had no horses, his other dog was quite inferior, and Tloght was the most valuable of his possessions. It will not then be wondered, that the story of this misfortune to the gallant Tloght, should have been the first intelligence communicated to me by the simple savage; or that he should dwell with a melancholy eloquence, on his virtues, and abilities, and

deplore his calamity. I sympathised with him, and made him a present of an axe and a kettle, for which he expressed his thanks in terms of warmest gratitude; and we parted, mutually pleased with the interview. I never saw him again.
On the thirteenth I left Flatt head post in a barge loaded with about a ton of merchandise, for the Horse plain, and manned by four stout Canadians; who propelled it with poles where the water was shallow, but when its depth would not admit of this mode of locomotion, recourse was had to paddles. We halted at sundown, opposite to a rock called "Le Gros Rocker". By noon on the following day, we reached House Prairie and encamped with a few lodges of Indians, who were awaiting the arrival of Mr. Ermatinger. In the afternoon, a Canadian reached us from Bitter Root river; he informed me that my fellow trader of last summer, was now with the Flatt heads. Mr. Ermatinger came up in the evening by land, with a quantity of goods upon pack horses. From this period until the twenty-third; Mr. Montour and myself, having purchased an equipment from the Hudson Bay Company; were employed in arranging bales, purchasing provisions, and making preparations for our departure to the Flatt-heads. We crossed over a point of the mountain, - around which the river winds, making a large curve in its course; - to the opposite side of the bend, following the same route, on which we had passed in December last; which we reached on the evening of the twenty-third. Leaving the Flatt-heads, we ascended a small branch, called Wild Horse Creek, to Cammas prairie; crossed over the mountain, and halted on the Arrowstone river, on the twenty-ninth. During our journey, we saw wild horses gallopping in bands over the plains, almost daily; several of which, were caught by our Indians and domesticated, with but little trouble. They pursued them, on very fleet horses until sufficiently near to "leash" them; when thus captured, they exert all their remaining force in fruitless endeavors to escape; and finally become gentle from exhaustion. In this situation they are bridled, mounted, and then, whipped to action. Other horses are usually rode before, that they may be induced to follow. If then they move forward gently, they are caressed by the rider; but on the contrary, most cruelly beaten if they refuse to proceed, or act otherwise unruly; a few day's practice seldom fails to render them quite docile and obedient. The process of catching wild horses, by throwing a noose over the head, is here called "leashing," and all Indians in the mountains, as well as those who rove in the plains east of them, are quite expert at it; although in this respect, far behind the inhabitants of New Mexico, who not only catch wild horses, cattle, buffalo, and bears; but even leash them by the feet, when at full speed, so as to render them quite incapable of moving. However, two experienced "leashers" are requisite to the complete capture of a large bull bison, or a full grown grizzly bear, and in both cases the feat is attended with considerable danger.
On our arrival at the Arrow stone river, we found it too high for fording, and immediately commenced making rafts. In the meantime, the squaws sewed up all torn places or holes in their lodges, then conveyed them, and their baggage, to the brink of the stream. The lodges were next spread on the ground and folded once in the middle, the baggage of several families were then placed on one,

taking care to put the heaviest articles at the bottom; the lodge was then firmly drawn from every side together at the top, and there strongly fastened by a thong, or rope, so that the whole appeared like a large ball four or five feet in diameter. About half way up the side two long cords or "leashes" (such as they use in catching horses,) were attached, and the ball then launched into the water, which buoyed it up like an egg-shell; several squaws and children then embarked upon it, and secured themselves from falling, by clinging to the cord which held it at the top. After all these arrangements were completed, two naked mounted Indians, seized the long cords between their teeth, and pushed out into the river; making their horses swim and carry them and tow one of their families and baggage at the same time. When I observed them start, I was fearful that the ball would upset and endanger the lives of the women and children; but was agreeably disappointed seeing them turning and returning, as the cords slackened or strained on it, horizontally; until it reached the opposite shore in safety. The river appeared in a few moments literally covered with these balls, all in the same manner constructed; all surmounted by women and children, and each towed over by two Indians on horseback. In short, they all crossed without accident, and transported our baggage over at our request, we having found our rafts quite insufficient. Ermatinger and myself, however, with several of his men, passed over on a raft; but the velocity of the current carried us down a considerable distance with fury, and it was for some time doubtful, where we should be able to effect a landing, but we finally passed so near a point of willows, which overhung the river, that we succeeded in arresting our rapid course by clinging to them, and got to shore, a mile below the landing place of the Indians, at which we all encamped.

Chapter XLIX

On the thirtieth, Mr. Ermatinger and myself, having learned that a party of Flatt-heads were encamped on Bitter-root river some distance above the mouth, set off in a gallop to see them. We arrived about noon, and had hardly gone through the tedious ceremony of shaking hands with them, before a large party arrived from Buffalo, with my old comrade Newell and several "engages" in company. From him, I ascertained that Dripps had received last fall an express from Fontenelle; stating that two of his men had been killed during the fall hunt, supposed by the Blackfeet, that Dripps had passed the winter at the forks, where I left him last November, found buffalo as numerous as during the preceding winter, and had met with no accidents. Late in the afternoon our party arrived, and we all concluded to remain here some time to recruit our horses.
In the evening of the third of May, while a party of the Indians were amusing themselves at a war dance, one of them, a spectator, carelessly resting his chin on the muzzle of his gun, was instantly killed by the unexpected discharge of its contents into his brain. The gun was probably, accidentally exploded by the foot

or knee of some person passing in the crowd. On the day following, the corpse was carefully dressed with a clean shirt, and blanket, then enveloped from head to foot, like an Egyptian Mummy, with robes and skins, well lashed around him, and finally committed to the silent keeping of the grave.

At length, our horses having become in good travelling condition, Mr. Montour and myself determined to set out in search of Americans, with whom we had resolved to trade; and started accordingly, on the tenth. Our party was composed of Mr. Montour, his son, two "engages," two Nez-perces, four Pend'orielles, one Cotena, one boy, four women and two or three small children. From the number of dialect in our camp, I am convinced, that a stranger would have been greatly puzzled to determine, which of the five languages continually spoken there, was predominant. However, we understood each other sufficiently well, to prevent mistakes; and the Indians comprehended one another, though they cannot be induced to convey their ideas in any tongue except their own. This custom would in a great measure prevent a proper understanding in many instances, were it not for their numerous signs, which constitute a kind of universal communication, not to say language, at once understood by all the various Indians in the mountains. These signs are made with their hands or fingers, in different positions, with rapidity; and are so extremely simple, that a person entirely unacquainted with them, will readily conceive a great portion of what may be expressed by them.

We passed, nearly southward, up the river some distance to a high point of mountain - projecting in the plain quite to the river's margin. We passed the Sheep Horn, which appears to have been garnished lately, with arrows, feathers, beads, pieces of scarlet, etc. in addition to the ornaments observed there in the fall. These little offerings, the superstitious natives ignorantly imagine will conciliate or gratify, the Deity, whom they suppose to have placed the horn in that singular position, that it may remain an eternal monument of his kind care for them, by continually reminding them, that it is their constant duty to remember and adore him. A confession or acknowledgement of their faith is made manifest, by the contribution of some little trinket or other trifling present. We crossed over the point, and descended into a fine little valley, inhabited by an extensive village of prairie dogs, where we finally halted on the second day about sunset. On the twelfth we crossed the mountain without difficulty, and encamped in the edge of the Big Hole. We saw several forts, made by the Black-feet since we passed here last fall. Next morning the last of our provisions, amounting to but a small piece for each person, was eaten for breakfast. We hoped, however, to kill something soon, and followed the Indian trail across this hole to Wisdom river, on which we halted for a short time, to bait our horses. It commenced raining soon after we started, and continued all day. Notwithstanding the storm however, we set out again, and proceeded to the warm springy where we encamped. The Indians expressed a desire to return back to the village. We were all cold, hungry and wet, and were but scantily supplied with sage roots and small willows for fuel. Not a few faces were chattering like rattle-boxes, and the prospect seemed fair for an uncomfortable night. A dram of Demarara rum, and a pint of hot coffee to each individual, was made to answer for fire and supper, and a night's meditation in a cold wet blanket for lodging, though they were considered by some as rather poor apologies for these common comforts. There was no choice, however; and during the night the drowsy phantom, continually flitting about like an "ignis fatuus," though often near, invariably eluded our grasp, as we essayed to catch it, and having spent the whole night in such fruitless endeavors, we arose, and collected our horses, who had broken loose from their pickets during the night, and rambled down the plain several miles. Having breakfasted, as we last night supposed, we set out, notwithstanding the storm, which still continued with violence; leaving the Big Hole behind us, we bore eastward over high bold hills into the Little Hole, which we followed, gradually turning to the southward, until near night, when we halted at its southern extremity. Our Indians killed during the march a cow that was poor and unpalatable; but were however, more fortunate in the evening, when they brought in the meat of a bull in much better condition. On the morning of the fifteenth we set out in fine spirits, to which a good night's rest, a full stomach, and a fair pleasant morning had severally contributed; and passed over a rising plain south-eastward, through Horse prairie, and halted for the night on a small stream, that flows into the eastern extremity of this plain. We saw during the day large herds of buffalo, killed several in fine eating order, and fared "pretty well considering," as a brother yankee would say, after a good meal. On the succeeding day we passed a high conical mound or spur, which it is, of the mountain, that is visible from any part of Horse prairie, and halted in the evening on the head of a creek, in an open valley, called by the Indians "Sin ko lats sin la," or "Hanged Man's Prairie," from the circumstance of one having been punished in that mode here by his fellows, for some crime several years since. We made that day twenty-five miles, our course south-east.

Chapter L

On the seventeenth, we ascended a bold hill, and came in view of the plains of Snake river, and the "Trois Titons", which bore nearly East; we descended the rough irregular plain, and halted at evening on the waters of the Columbia. The face of the country here is extremely broken, and intersected by small streams, that are separated from each other by high bluff rocky hills. We saw immense herds of buffalo in the plains below, which were covered in every direction by them, as far as the eye could distinguish. We were cautioned by our Indians during the march, to prevent our horses feeding by the way, in consequence of a poisonous herb, resembling a parsnip, which abounds there, and causes death shortly after being eaten.
On the eighteenth, we travelled South-Eastward down to the great plains, and halted on Poison Weed Creek. We observed columns of sand moving in various

directions; raised by whirlwinds, though the day was fair and the air still, without a breeze or even zephyr. These singular columns, when passing over the plain with great rapidity, would often suddenly stop, as if to gather more force and a denser cloud of dust, for some moments, revolving round with accelerated velocity; then again progressing slowly, move off in quite a different direction from that in which they run, before halting; with faster flight they would wing their way along, until again they would pause at some sand hill, to increase their speed or condense their substance, and again pass away with volume and vigor apparently renewed; until they would finally dissolve and disappear. The most extensive and remarkable of these whirling pillars, were seen to rise on the hills, at the base of the Sand Mountain, and large quantities of the beautiful fine white sand of which they are composed, were scattered over the plain by these aerial phenomena.

I had heard in the summer of 1833, while at rendezvous, that remarkable boiling springs had been discovered, on the sources of the Madison, by a party of trappers in their spring hunt; of which the accounts they gave, were so very astonishing, that I determined to examine them myself, before recording their descriptions, though I had the united testimony of more than twenty men on the subject, who all declared they saw them, and that they really were, as extensive and remarkable as they had been described. Having now an opportunity of paying them a visit, and as another or a better might not soon occur, I parted with the company after supper, and, taking with me two Pen-d'orielles, set out at a round pace, the night being clear and comfortable. We proceeded over the plain about twenty miles, and halted until daylight on a fine spring, flowing into Cammas Creek. Refreshed by a few hours sleep, we started again after a hasty breakfast, and entered a very extensive forest called the Piny Woods, which we passed through, and reached the vicinity of the springs about dark, having seen several small lakes or ponds, on the sources of the Madison; and rode about forty miles; which was a hard day's ride, taking into consideration the rough irregularity of the country through which we had travelled.

We regaled ourselves with a cup of coffee, and immediately after supper lay down to rest, sleepy, and much fatigued. The continual roaring of the springs, however, for some time prevented my going to sleep, and excited an impatient curiosity to examine them; which I was obliged to defer the gratification of, until morning; and filled my slumbers with visions of water spouts, cataracts, fountains, jets d'eau of immense dimensions, etc. etc.

When I arose in the morning, clouds of vapor seemed like a dense fog to overhang the springs, from which frequent reports or explosions of different loudness, constantly assailed our ears. I immediately proceeded to inspect them, and might have exclaimed with the Queen of Sheba, when their full reality of dimensions and novelty burst upon my view, "The half was not told me." From the surface of a rocky plain or table, burst forth columns of water, of various dimensions, projected high in the air, accompanied by loud explosions, and sulphurous vapors, which were highly disagreeable to the smell. The rock from which these springs burst forth, was calcareous, and probably extends some distance from them, beneath the soil. The largest of these wonderful fountains, projects a column of boiling water several feet in diameter, to the height of more than one hundred and fifty feet - in my opinion; but the party of Alvarez, who discovered it, persist in declaring that it could not be less than four times that distance in height - accompanied with a tremendous noise. These explosions and discharges occur at intervals of about two hours. After having witnessed three of them, I ventured near enough to put my hand into the water of its basin, but withdrew it instantly, for the heat of the water in this immense cauldron, was altogether too great for comfort, and the agitation of the water, the disagreeable effluvium continually exuding, and the hollow unearthly rumbling under the rock on which I stood, so ill accorded with my notions of personal safety, that I reheated back precipitately to a respectful distance. The Indians who were with me, were quite appalled, and could not by any means be induced to approach them. They seemed astonished at my presumption in advancing up to the large one, and when I safely returned, congratulated me on my "narrow escape." - They believed them to be supernatural, and supposed them to be the production of the Evil Spirit. One of them remarked that hell, of which he had heard from the whites, must be in that vicinity. The diameter of the basin into which the water of the largest jet principally falls, and from the centre of which, through a hole in the rock of about nine or ten feet in diameter, the water spouts up as above related, may be about thirty feet. - There are many other smaller fountains, that did not throw their waters up so high, but occurred at shorter intervals. In some instances, the volumes were projected obliquely upwards, and fell into the neighboring fountains or on the rock or prairie. But their ascent was generally perpendicular, falling in and about their own basins or apertures. These wonderful productions of nature, are situated near the centre of a small valley, surrounded by pine covered hills, through which a small fork of the Madison flows. Highly gratified with my visit to these formidable and magnificent fountains, jets, or springs, whichever the reader may please to call them, I set out after dinner to rejoin my companions. Again we crossed the Piny Woods, and encamped on the plains at Henry's fork.

Chapter LI

Resuming our journey early on the morning of the twenty-first, we crossed Henry's Fork, and continued on till we arrived in the vicinity of Pierre's Hole; where the country assumes a rolling appearance, and is dotted with an occasional grove of aspen trees. Entering this valley, we passed over to Pierre's fork, near which we discovered the trail of our company, and following it about three miles, at our rapid pace, overtook them as they were on the very point of encamping. Mutual congratulations over, I retired much fatigued to rest. From Mr. Montour, I learned that the hunters had killed several bulls, a bald eagle and a goose. They were greatly annoyed by suffocating clouds of dust, which arose from their horses' feet, filled their lungs and eyes, and the air around them, for some distance.

On the twenty-second, we remained to dry meat, which was prepared and packed, ready for transportation, by evening. Next morning we pursued our journey, passed South-Eastward to Pierre's Hole, and halted in the mountain, on a trail leading over it. We killed, during our march, five buffalos and an antelope. One of our Indians found in Pierre's Hole, a pair of boots, and some articles of clothing, that had evidently been there a long time, on the prairie; probably lost by some white man during last year. On the twenty-fourth we ascended the mountain, crossed an immense snow bank with extreme difficulty, descended the plain, passed through Lewis's rim, though so high that our horses were obliged to swim, and halted on a small stream several miles east of the river. Leaving the river on the twenty-fifth, we passed along this valley to the South-East point or extremity, when we reached a fork of this stream, and travelled up its narrow bottom, flanked on either side, at the distance of less than half a mile, by lofty mountains; climbing occasional hills or bluffs, which project in some instances quite to the river's margin; and halted at the commencement of the narrows, formed by the mountains closing upon the stream, until barely sufficient space remains for its compressed channel. We here found an encampment, made by a large party of whites, some ten days since, on their way from Salt river to Green river; I supposed it to have been made by Dripps, who wintered on Snake river. On the twenty-fifth we passed, with our usual hazzards and difficulty, though fortunately without accident, through the tortuous windings, abrupt elevations, and percipitous descents, of the Narrows, out of which we were glad to emerge; and entering Jackson's Little Hole, encamped on a small branch of this fork, at the East side of the valley. Antelopes and buffalos were found here; and an encampment made by the company whose traces we observed the day previous. On the twenty-seventh we ascended the steep, rough, aspen covered hill, forming the east boundary of this hole, and passing down on the opposite side, came into the plains of Green river. We now directed our course towards "Bonnyville's Folly," or "Fort Nonsense," as it was more frequently called; but had proceeded a few miles only, when we discovered two Indians so near us that they could not hope to escape though they betrayed considerable anxiety at our approach. When we reached them, however, their fears were quelled, we learned that they belonged to the party of Dripps, who were encamped on a small stream in the Wind mountain, east of Green river. On hearing this information, we turned our course to that stream, which we crossed without accident; and halted after sundown in the willows, on the border of a small branch; after a hard march of thirty miles, at least. Early the next morning we set out, found the trail of Dripp's party, followed it ten miles to a small stream, and then found an encampment, which had been that morning evacuated. We continued to pursue the trail, over a rough, rocky, hilly country, and finally descended to the margin of a fine little lake, about ten miles long and one broad; down which, the trail passed to the western side, and finally conducted us to Dripps' encampment, in a narrow bottom, bounded on one side by a high, rocky, bold hill; and by the lake on the other; our course here, from Green river, was nearly East.

From Mr. Dripps we learned, that Fontenelle and others were in the Eutaw country, to the Southward, trapping still. This lake, fed by springs, constitutes the source of the Western branch, of the New Fork. The trappers informed me that there were several other small lakes in this side of the mountain, on the sources of the other branches of this fork. We remained here for several days, awaiting the return of some small parties of trappers. On the thirty-first, we made a short march to the outlet of this lake, for better grass, and killed on our way several buffalos. On the succeeding day, June the first, we collected all our horses together at camp, and blooded them indiscriminately, to make them thrive, and render them more healthy. This operation is performed every spring, and is considered quite necessary. Indeed horses who are bled, invariably fatted faster, and become strong, hardy, and active, much sooner than those who are not. June, the second, a party of trappers returned, who had been out since last fall; several of them saw Fontenelle this spring, in the neighborhood of Bear river. He had lost one man, killed, by the name of CHEVALIA. We moved several times, short distances parallel with Green river, and finally went to it. In the mean time, an express was despatched to Fontenelle, and several small parties of trappers, returned from various sections of the country, and related their adventures and escapes. The weather had been pleasant, we seldom failed to kill plenty of buffalos when we made an effort to do so, and we therefore fared and frolicked, when we chose, agreeably.

Chapter LII

Several men bitten by a rabid wolf - Arrival in camp of two Eutaw squaws from the Snakes - Wild currants, goosberries, etc. - Mexican Indians

About this time we learned that two persons, who were bitten by a wolf, at last rendezvous, had died or disappeared suddenly. The circumstances during the hurry and bustle of business at rendezvous were by mistake not recorded in my journal, though they produced great excitement at that time. They were as follows: whilst we were all asleep, one night, an animal, supposed to be a dog, passed through camp, bit several persons as they lay, and then disappeared. On the following morning considerable anxiety was manifested by those who were bitten, under the apprehension that the animal might have been afflicted with the hydrophobia, and several of them took their guns and went about camp, shooting all suspicious looking dogs; but were unable to determine that any one was positively mad. During the day information came from the R. M. F. Co., who were encamped a short distance below us on the same side of the river, that several men were likewise bitten in their camp during the night, and that a wolf supposed to be rabid, had been killed in the morning. The excitement which this affair originated, however, gradually subsided, and nothing more was heard of mad-dogs or wolves. In the fall subsequent, one the persons who had been bitten, a young Indian brought from the council Bluffs by Mr. Fontenelle, after having given indications of the hydrophobia, disappeared one night from camp and was heard of no more. The general impression being, that he wandered off while under its influence, and perished. Another individual died of that horrible malady, after having several violent spasms, while on his way from the mountains to St. Louis, in company with two others. Whether there have been any more instances of the kind, I am not informed.

On the twenty-fourth, two women arrived at our camp, in a starving condition. One of them had an infant at her breast, and was the wife of Cou-mar-ra-nap, a famous Eutaw chieftain; the other was a young, unmarried girl, of the same nation. They were taken prisoners by a party of Snakes, during the early part of the summer, and were conducted to the village, where they were condemned to die; but were saved by the timely interference of Capt. Walker, who humanely purchased their lives, and sent them to overtake us, as the surest means of getting to their own country in safety. They had followed our camp twenty days, living upon roots and berries, and had avoided the trails and most frequented places, for fear of again falling into the hands of their enemies. The Snakes declared war against the Eutaws last fall, for killing several of their tribe, who were caught in the act of stealing their horses. A few days after capturing the women, they stole the horses of a party of hunters, from the Rio del Norte, on a stream called the Euruta, and returned to their village with the booty.

On the twenty-ninth we passed southward over a rolling country, covered with cedars; and halted at a small stream, that discharges itself into Green river. The borders of this creek were covered with bushes, laden with fruit, which was now ripe. There were black, white, yellow and red currants, large as cherries; though to the taste, quite inferior to the common garden currants, being less sweet, and more acid; plump goosberries of a large size, likewise tempted the eye, but were equally sour. The bushes that bear them, frequently attain the height of eight or ten feet, in the rich mellow soil along the rivers; and at this season, are bent to the ground by the loads of fruit, with which they are encumbered. There was also a species of small tree, not unlike the hawthorn, armed with thorns, and covered with blue or lead colored leaves, which was completely enveloped with berries, either red or yellow, about the size of allspice or pepper grains. This fruit is extremely sour, and is commonly called "buffalo berries." They are prized by the Indians for abundance, surpassing by far, all other productions of this kind, and quantities of them are collected by the squaws for food. It may be remarked, in relation to goosberries, that we have frequently found them in other parts of the mountains of a delicious flavor, as well as of a very large size.

On the third of September, being within the territory of Mexico, three of the wild natives of this region, ventured into our camp. They were stark naked, and betrayed almost a total want of intellect, which was perhaps the result of extreme wretchedness and misery, to which they are continually exposed, from infancy until death. They spoke a tongue perfectly unintelligible to us, and evidently not well comprehended by each other; neither could they understand any of those expressive signs, and gestures, by which other Indians convey their ideas, with perfect success. Severe hunger, however, instinctively taught them to make us understand, that they wanted something to eat, which indeed was all they could communicate. We gave them the carcases of two beavers, together with the head and feet of a dog, which they warmed, a few moments, over the fire, and devoured with wolfish avidity. A Nezperce with our party, regarded them attentively, until they had finished their repast, when he arose, and with evident marks of astonishment, in his countenance, exclaimed "why my horse has got more sense, than those Indians!" It may be well to remark in this place, that the different nations in the Rocky Mountains are, for the most part, confined to certain districts, called by them, their own country; within which they rove, and become familiar with every trifling object; but are almost entirely ignorant of the country, or people, beyond their limits. The Indian who made the remark above, was born on one of the sources of the Columbia, but was not aware of the existence of this nation, until we brought him among them.

On the fourth, four of these Indians, who call themselves "Sann-pitch," came into camp, bringing to my surprise, several deer skins. I say, to my surprise, because from the hasty opinion I had formed yesterday, I imagined deer, to be the more intelligent animal of the two, and should have expected them with skins of the Sann-pitch, marching up to trade, almost as soon. However, in justice to the superiority of the latter, skins they brought, and made us understand, by simply pointing to the skins, and then toward our horses, that they wanted one of them. We quickly informed them, that our horses could not be disposed off for deer skins. They manifested a great deal of disappointment, and finally offered them for provisions, but were equally unsuccessful. These are by far, the most miserable human beings we have ever seen. The barreness of their country, and scarcity of game, compel them to live by separate families, either in the mountains, on in the plains. In the latter, they usually select the most barren places to encamp, where there is apparently nothing but sand, and wormwood or sage. Here, the women and children are employed in gathering grasshoppers, crickets, ants, and various other species of insects, which are carefully preserved for food, together with roots, and grass seed. From the mountains, they bring the nuts which are found in the cores of the pine, acorns from the dwarf oaks, as well as the different kinds of berries, and the inner bark of the pine, which has a sweet acid taste, not unlike lemon syrup. In the mean time, the men are actively employed in hunting small animals, such as prairie dogs, squirrels, field mice, and larger animals, or birds, which fortune some times, places within the reach of their arrows. They likewise take fish, with simple instruments of their own invention, which will be hereafter described.

The Sann-pitch are generally quite naked, though in some instances a small piece of skin, is fastened before them. The women all wear a piece of skin, reaching from the middle, to the knees, and instances are not uncommon, where they possess a complete leathern shirt, but no other article of dress. They are extremely shy, and approach us with evident fear and caution. If in the plains, they conceal themselves from our approach, by crawling into the sage, or into gullies; but if discovered in the open prairie, where flight would be useless, they throw themselves flat upon the ground, in hopes of being mistaken for a rock, or other unusual appearance; which practice generally succeeds, if they were not discovered before putting it into effect.

Chapter LIII

A bed of salt - Indian arrows - Brutal conduct of a hunter towards an aged Indian - Chanion of White River.

On the eight, I set out with others to procure salt, at a place discovered by our hunters yesterday. We passed three miles down the river, and found the salt in a slough on the west side of it. It was found on the surface of a black stinking mire, fifty or sixty paces in circuit; the upper strata was fine, and white as snow, to the depth of two inches; beneath which, was a layer of beautiful chrystals, to the depth of five or six inches, that rested on the surface of the mire. We slowly sank into the latter to our knees, whilst scooping up the salt, and then changed places, for we could scarcely extricate ourselves at that depth; and concluded that if we should remain long enough in the same spot, we would at length disappear entirely. This opinion was coroborated by thrusting down a stick four feet in length, without meeting any resistance, more than at the surface. I gathered about a half bushel in a few minutes, and returned with my companions, who were equally fortunate, to camp.

*****

I observed during our stay on the Sararah, that the Indians had two kinds of arrows in their quivers, one of which was made of a single hard stick, feathered and pointed with transparent flint, artfully broken to a proper shape, and firmly fastened to the end of the arrow with sinews and glue. The others were made of a hollow weed, having six or eight inches of hard wood nicely inserted, and firmly glued into it; to the end of which the stone point is fastened, and is poisoned with venom from the fangs of a rattle snake. Hence the slightest wound from them is certain death. These arrows may be known at sight, by the natural joints of the cane; and the artificial one, where the wood part is inserted. They are not solely used in battle, as some have asserted; but are equally advantageous in hunting, for the slightest wound causes the animal to droop, and a few moments places it within the power of the hunter. The flesh of animals thus poisoned, is harmless in the stomach.
On the nineteenth, we continued northward, over a gently ascending plain, and encamped on a small stream, that flows into the Eutaw lake. During our march we encountered a feeble old Indian, whose age and infirmities, if they could not

have insured him respect, ought at least to have shielded him from harm. Innocent and inoffensive, this miserable old man - destitute of the means of offense, and engaged in the harmless occupation of gathering roots, his only apparent means of subsistence, which he deposited in a willow basket on his back - was overtaken by one of our heartless comrades, who, having had a valuable horse stolen on the seventeenth, in a most unfeeling manner, inflicted a severe blow upon his head, with his gun. For once the fortitude of the Indian yielding to the frailty of nature, he gave utterance to a scream of agony, which was distinctly heard by all of us, though far in advance of them. I galloped back as soon as possible, and saw him covered with blood, while a few paces distant stood the hunter in the act of loading his gun, with the avowed intention of taking his life. I easily persuaded him, however, to leave the Indian without further molestation, and he accordingly proceeded with me to the company, heartily ashamed of his brutal conduct. Similar instances I am happy to say, are of rare occurrence.

On the twenty-ninth, we entered a narrow passage between two formidable walls of cut rocks, called by the hunters the Chanion of White river; which rose, perhaps, from one to two hundred feet in perpendicular height, and sixty or eighty yards asunder. This narrow space is chiefly occupied by the river, winding from one side to the other, as if enraged, at being thus confined. The walls are seldom accessible, and are surmounted in some places, by singular peaks of weather-worn sand stone resembling, when beheld at a distance, domes, turrets, steeples and towers, so strikingly that a single glance is sufficient to excite in the mind of the spectator, the idea of a flourishing village, and the vicinity of a civilized country. But, alas! a nearer and more careful view, dispels the pleasing illusion, changes those spires to solitary desolation, and turns the lovely creations of imagination, to naked cliffs and sandy deserts, far, far from the inspiring presence of home, and the affectionate relations of social life.

Chapter LIV

Arrival to camp of twenty families of Eutaws - An Indian Chief - Several horses stolen from the camp, and attendant circumstances - Shaving materials, etc. found - Murder of a squaw by her husband.

On the first of November, we were joined by twenty families of Eutaw Indians, who were returning from Buffalo, having loaded their horses with dried meat. They had several stolen, whilst they were engaged in preparing their meat, for transportation, which induced them to retreat expeditiously, fearing an attack from their enemies, the Snakes. This was effected with so much rapidity, that they packed their horses unskillfully, and nearly ruined them; for, in consequence of such neglect, many of them had their backs completely skinned. A party of Indians also came in from the southward on the fourth, and we immediately opened a brisk trade with them for furs, deer skins, etc. The principal chief is a hardy warrior, about forty years of age; evidently superior both in a mental and physical point of view, to any of his followers. His hair, which is of uncommon length, he wears coiled up in a knot on his forehead, secured by a thong; but differs from his tribe in no other particular, either of dress or ornament. The expression of his countenance is mild and thoughtful, and rather pleasing than otherwise. His keen wandering eye bespeaks intelligence, and his demeanor is dignified and impressive. He is reputed the bravest man of his nation; and has won, on the sanguinary field more trophies, than any of his warriors. He is known to the hunters by the name La Toque, which he has received from our French comrades, probably because his hair, has some resemblance to a cap. His followers do not steal, but are continually loitering about our encampment, curiously examining every thing they see, and never fail to ask for every thing they examine; hence they are even more intolerable, than some of their less honest neighbors.

On the tenth, we were alarmed early in the morning by the cry of "robbers," and immediately sallied out to collect our horses, - twenty of which belonging to the Indians were missing. We always fastened ours in camp at night, whilst the Indians frequently permitted theirs to run loose, and were consequently the sufferers. Unfortunately, one of our men had discovered maces of Indians, some distance below our encampment last night; but neglected to inform the chief, and thereby put him on his guard. The consequence of this neglect on our part, was a universal belief among the Indians that the robbers were Snakes, with whom they knew we were friendly, and that we had seen and given them the requisite information, to the accomplishment of their design. No arguments of ours could prevail against this opinion, which was based upon a circumstance that occurred last fall in the camp of Fallen and Vanderburgh. They were encamped at that time on a river thirty miles north of us, and had been accompanied by a family of Eutaw Indians, during their hunt, who intended to remain with them until spring. But on reaching that river they were joined by a party of Snakes, who watching a favorable opportunity, decoyed a young man of the Eutaw family from camp, and slew him. The whites whose interest taught them to remain friendly with both nations, refused to tend their aid to avenge the deed; and the exasperated father, with his wife, and children, left them the same night; but was pursued and overtaken by the Snakes, who massacred them all. The particulars of this murder the Eutaws heard from ourselves, but blaming the whites at large for the fault of a few, in permitting so flagrant an outrage on the rights of hospitality, to pass unpunished, they could view us in no other light, than as accomplices. Hence they doubted not, that we had formed an alliance with their enemies, to effect their destruction. Fully persuaded of this, the chief instructed the women to saddle their horses, and be ready to fly into the mountains, should his fears be confirmed; whilst himself and many of his boldest followers, set out in pursuit of the robbers. Several of our men, who wished to prove their suspicions groundless, accompanied them; but returned in the evening with some of the Indians, having continued the chase to Bear river, thirty miles to the northward;

where the course taken by the robbers proved them to be Arrappahoes, with whom we were at war as well as the Eutaws. This fact immediately restored confidence between us, and our men having no other object in view, returned to camp; but twenty of the Indians still followed on.

On the seventeenth, some of the Indians returned from hunting with a razor, shaving box, two shirts, and a blanket, which they found near a small stream. They had lain exposed to the weather for months. But of their being found here, there were no marks on either of these articles, by which the mystery could be unravelled.

On the evening of the twenty-seventh, one of the Indians shot down his wife in a paroxysm of passion, for some trifling misdemeanor, and having thus satiated his demoniac rage, fled. The squaws immediately joined in a heart rending lamentation, and the men rushed out of their lodges, to avenge the deed. For some minutes the camp presented a scene of clamor and confusion, that bade fair to end still more seriously. The lodge of the murderer disappeared in a moment, his property was quickly destroyed by the enraged friends of the deceased, and his horses would have shared the same fate, had they been in the care of one less resolute than the chief, who stood beside them gun in hand, and forbade their approach on the peril of their lives. No one dared to oppose the haughty chieftain, who had set them all at defiance, and the outcry soon subsided, though had the wretch who perpetrated the deed, fallen into their hands, I am confident he would have been torn to pieces in a moment.

Chapter LV

Conmarrowap, a noted Chief of the Py-Euts.

On the twenty-third, Conmarrowap, a celebrated Chief, with his wife, and ten warriors, came into camp. This noted personage is a Eutaw by birth, but forsook his own people and joined the Py-Euts, after he became a man, and by his prowess and bravery, acquired such an ascendency over the tribe of his adoption, as to become their principal chief. He has rendered himself as an object of terror to them, by an atrocious custom of taking their lives, for the most trivial offenses.. He is the subject of conversation every where among the Eutaws, by whom he is universally detested; all agreeing, that he deserves death, but none can be found daring enough to attempt its accomplishment.

He is the only Indian in the country, who ever dared to chastise a white man, in his own camp; and had not the partisans of the hunter interfered, his soul at that time would have taken its flight to eternity; for the high spirited trapper could not brook from the haughty Chieftain, an insult, that would have awakened a spirit of vengeance in the breast of the meanest Indian, and immediately leveled his rifle at the heart of his intended victim, who, perhaps for the first time in his life, betrayed emotions of fear. However, the comrades of the justly exasperated hunter, dashed his gun aside, and prevented the execution of a deed which would certainly have been avenged by the Indians; though the poor trapper shed tears of regret for the loss of an opportunity of appeasing his wounded honor, and at the same time punishing a savage tyrant. This evil genius once fell in with a party of trappers, some of whom are now present, and bored his finger in their ears, to make them more readily comprehend him; but I doubt whether this treatment rendered their understandings more susceptible to Eut -gibberage afterward. I heard one of these Indians seriously ask a hunter, with whom he was conversing, to allow him to spit in his ear, assuring him that if he would permit it, he must inevitably understand the Eut language thence forward; but he seemed more inclined to laugh at the folly of the proposition than to submit to the mode of instruction.

Conmarrowap's wife and her companion, after leaving us last Summer, fell in with the relations of the latter, who unhesitatingly killed and devoured the horse, we had given to the former. Leaving the inhospitable relations of her companion, she proceded on, and reached her husband some days after. He had been sick during several weeks, and for some time was considered past recovery; but survived, and as soon as sufficient strength returned, set out to visit those who had robbed his wife. An altercation ensued, which resulted in the death of the man who was at the head of the family that had injured him though not until he had received a slight wound from an arrow himself. He lost all his horses last summer, when his wife was taken prisoner, yet he now had ten of the finest we have ever seen among the Indians. He says they were presented to him, by the passing traders from Toas to California; but it is much more probable that he took them by force, as he has already done to our knowledge in many instances. All the hunting parties from Toas look upon him as a terrible fellow, and submit to his insults, which they dare not resent; although I have seen one or two individuals, who have sworn to take his life, the first opportunity that occurs, when they may not endanger themselves. During our stay on this river, one of the log huts was occupied by those trappers from Toas, who joined us last fall. They brought with them from the Sources of the Eurata, the flesh of several fine deer, which was piled upon some branches, in a corner of the room. Conmarrowap's wife entered the house one day and asked for some meat; but as the men did not immediately attend to her request, she departed without any; in a few moments Conmarrowap himself hastily entered, and without speaking, applied his herculian strength to tossing the meat to and fro, until he found the best deer in the lot, which he shouldered and with it disappeared; whilst those men sat in mute astonishment, nor dared to cast a glance of dissatisfaction towards him; though he was armed with nothing but a frown, from which they shrunk back with awe, having before felt the consequences of his displeasure. There is nothing uncommon in the appearance of this Indian, save a stern and determined look; he is now slender, of a middle stature, and has a dark, keen and restless eye; but before his sickness, was quite corpulent, a rare circumstance among Indians. There is less in his dress and manners, to distinguish him from his fellows, on ordinary occasions. He appears to be about forty years of age.

Chapter LVI

A Winter Encampment

The season having become far advanced, we pitched quarters in a large grove of aspen trees, at the brink of an excellent spring that supplied us with the purest water, and resolved to pass the winter here. Our hunters made daily excursions in the mountains, by which we were half surrounded, and always returned with the flesh of several black tail deer; an animal almost as numerous as the pines and cedars among which they were found. They frequently killed seven or eight individually, in the course of a day; and consequently our encampment, or at least the trees within it, were soon decorated with several thousand pounds of venison. We passed the time by visiting, feasting, and chatting with each other, or by hunting occasionally, for exercise and amusement. Our camp presented eight leathern lodges, and two constructed of poles covered with cane grass, which grows in dense patches to the height of eight or ten feet, along the river. They were all completely sheltered from the wind by the surrounding trees. Within, the bottoms were covered with reeds, upon which our blankets and robes were spread, leaving a small place in the centre for the fire. Our baggage was placed around at the bottom of the lodge, on the inside, to exclude the cold from beneath it, and each one of the inmates had his own particular place assigned him. One who has never lived in a lodge, would scarcely think it possible for seven or eight persons to pass a long winter agreeably, in a circular room, ten feet in diameter, having a considerable portion of it occupied by the fire in the centre; but could they see us seated around the fire, cross legged like Turks, upon our beds, each one employed in cleaning guns, repairing moccasins, smoking, and lolling at ease on our elbows, without interfering with each other, they would exclaim, Indeed they are as comfortable as they could wish to be! which is the case in reality. I moved from a lodge into a comfortable log house, but again returned to the lodge, which I found much more pleasant than the other. These convenient and portable dwellings, are partially transparent, and when closed at the wings above, which answer the double purpose of windows and chimneys, still admit sufficient light, to read the smallest print without inconvenience. At night a good fire of dry aspen wood, which burns clear without smoke, affording a brilliant light, obviates the necessity of using candles. Our little village numbers twenty-two men, nine women and twenty children; and a different language is spoken in every lodge, the women being of different nations, and the children invariably learn their mothers tongue before any other. There were ten distinct dialects spoken in our camp, each of which was the native idiom of one or more of us, though French was the language predominant among the men, and Flathead among the women; yet there were both males and females, who understood neither. One would imagine that where such a multiplicity of tongues are spoken, a confusion, little short of that of Babel, would naturally ensue. However, it is not the case.

Men who find it difficult to convey their ideas to each other, through ignorance of their opposing dialect, readily make themselves understood by avoiding difficult or abstract expressions, and accompanying their simple speech with explanatory gestures.

Chapter LVII

An attempt to pass through a frightful chasm - Battle offered to a grizzly bear - Profound cogitations of a grey wolf

On the thirtieth of March, we entered the chanion of a deep creek, and attempted to pass through one of the most frightful chasms, perhaps, in existence. On either side of the narrow space at the bottom, which was thirty paces in width, huge perpendicular walls arose, to the apparent height of a thousand feet, and were surmounted by large pines, which appeared but as twigs from the abyss beneath. We had not penetrated far beyond the entrance to this murky cave, whose gloomy vaults have probably never been explored by mortal footsteps; and whose recesses are veiled in a shroud of everlasting night, when we found it impossible to proceed further with our horses, in consequence of innumerable obstacles by which the passage was obstructed. Dismounting, I forced my way some distance over banks of snow, that entirely bridged the stream - huge fragments of immense rocks, which had been precipitated from the summit of the walls, and crumbled to pieces on the stony pavement below - lofty pines that had been torn up by descending avalanches, which accumulate on the steep side of the mountain, until even the gigantic trees can no longer sustain the superincumbent pressure of the mighty mass, that gliding down resistlessly sweeps all before it, over the eternal cliffs, and thundering thence with accelerated fury, mingles in one chaotic heap of shapeless ruins below. - Deterred from advancing further by the sublime terror of the surrounding scene, and the absolute impossibility of accomplishing my design, I stood for a few moments in mute astonishment and wonder - profusely showered by the ever-dropping spray from elevated regions of perpetual snow - beside a stream, whose foaming current has for ages dashed impetuously along its rock-bound channel - gazing in awe upon the dangers that had already stripped of branches a few lonely monarchs of the soil, that towered in solitary grandeur here and there, and seemed to mourn the dismal desolation scattered around and still threatened destruction. Turning my eyes aloft, the view was menacing indeed. Overhanging trees and rocks, and banks of snow, were poising on the verge of this dire descent, and on the very point of tumbling off: perhaps Imagination or dismay made them appear to tremble, but so they did. Shivering with dread, I turned from a place, so pregnant with dangers, and hastening forward, heard, as I returned, the deafening sound of timber, stones and snow, hurled down the giddy heights from crag to crag, into the gulph below. Rejoining my companions who had heard appalled those tremendous concussions, we gladly abandoned a spot so dreary, and proceeding to the plain halted on a portion of the prairie, that was covered with dry sticks; having been the camping ground of a party of Indians, a year or two since.

***

On the 14th April, having placed a rag at the extremity of a stick, planted in the ground near camp, to prevent the wolves from rifling it, we all set out to combat a grizzly bear that had buried the carcass of an animal, some distance up, near the margin of the river, and had put Blackface to flight last evening whilst heedlessly approaching his prey. We reached the spot after riding four miles, and opening a little mound of fresh earth, discovered and disinterred the entire carcass of a large elk, recently killed. We remained some time awaiting the appearance of the bear, but our numbers probably deterred him from leaving the thicket in which we supposed he was concealed, being but a few paces from his charge, and from which his traces led to and fro. However, we returned disappointed to camp, where every thing remained as we had left it, though a large grey wolf sat upon his haunches a short distance off, having evidently (as an Indian would express it) two hearts, for and against, helping himself to some of the fresh elk meat that lay exposed to view with perhaps for him the same attractions, that a roast pig would have had for one of us. His cogitations on the propriety or expediency of charging up to the luscious store, and committing larceny, in open defiance of the fearful banner waving over it, being interrupted by our approach, his thoughts immediately turned into a new channel, and he came at once to the sage conclusion, with wonderful alacrity and sagacity, that -

"He who coolly runs away,
May live to steal another day,"

and proceeded to put the decision into practice, without unnecessary delay.

Chapter LVIII

A party of hostile Indians out-generalled, etc.

About two hours before sunset, we observed a large party of Indians running about a mile in advance of us, evidently seeking a situation favorable to entrap us, by an ambush. They were scarcely perceptible, owing to the vapors rising from the damp plain; and as they had taken a position near which we must inevitably have passed, it was a lucky chance, or providence, that saved our lives by enabling us to anticipate their design. We halted apparently as if we had not discovered them, at a large cluster of willows, on a small stream flowing into

Bear river: we turned out our horses to feed, made a rousing fire, cooked and eat a hearty supper, and scattered our baggage carelessly about, to give our camp an unconcerned appearance, though every eye was bent on some point, ravine, or hill, watching the appearance of the enemy. - As soon as it became dark, we caught our horses, packed up, and moved off, with as little noise as possible. We discovered the Indians at the distance of a mile, east of us in a deep ravine, around a small fire, busily dodging about, evidently making preparations to steal our horses and attack us. We smiled to think of their disappointment, when they should find themselves outwitted, and that their expected prey had slipped through their fingers; and congratulated ourselves on the complete success of our stratagem. We laughed in our sleeves as we passed them, and chuckled with glee, as soon as we were out of danger, at the blank countenances they would exhibit at daylight, when they would find that we were really gone. The night was dark, and we pursued our way some time after midnight, guided by the polar star, from which, and a few other dim ones, what light there was, came twinkling down. We halted a short time, but daylight found us again moving, on the morning subsequent, on the highest part of the plain between Bear and Snake rivers; from which we discovered a large smoke near the latter stream, and supposing it to proceed from the camp of Vanderburgh, we continued on until we found ourselves in a Snake village. These Indians do not often kill whites in their own camps, though they have done so, in instances where they thought they would never be exposed. A man was at this time in one of our parties, who was wounded by and escaped from them. - As they are esteemed the most treacherous, cunning and vindictive Indians in the mountains, we did not feel ourselves safe with them, and continued our progress, satisfied that the party we saw last night belonged to this village. After enquiring where Vanderburgh's camp was situated, they denied having any knowledge of whites on this steam, but persisted in asserting that a company was yet on Bear river. We contradicted their statement, and pointed to the ground where the trail of Vanderburgh's party passed along near their village. They affected surprise, and declared that they had not before perceived it; though the lying, cunning rascals always examine every part they visit, and observe the most minute appearances, wherever they go. We followed the trail to Snake river, a few miles eastward, over several ranges of hills, but the country where he had promised to remain was covered with buffalos, and we concluded that he had gone to Green river; still we continued to follow the trace, passed through the timbered bottom, and reached the margin of the river at the ford, where we halted to await the arrival of one of our men, who had been allowed to go below, and ascertain if the company had left this river.

Shortly after we halted, about twenty Indians came rushing towards us, and yelling like so many devils. We seized our guns and sprang behind the trunk of a large fallen cottonwood tree; not doubting that the villains would attempt our lives. However the principal warrior, who was the ugliest looking man among them, dismounted, and made signs to us to come and smoke. We did so, and he enquired who were the two Indians with us; we made him understand that they were Delawares, because the Shawnees had fought and killed a party of several Snakes last summer, which they well knew; and to have acknowledged them Shawnees would have exposed them and ourselves to instant death; for they would have been butchered on the spot, and we were sworn to protect each other. We made them a small present, and they departed peaceably.

Shortly after they had left us, a hunter arrived from camp, which was situated only four or five miles below on the opposite side of the river, and we immediately packed up and forded the stream without accident, tho' the water was so deep as to come quite over the backs of our smallest horses; and went to camp, which we found situated in a fine timbered bottom, on the margin of the river. Nothing remarkable had occurred during my absence, and the trappers had all returned, except Williams and his companions.

Chapter LIX

Number of Hunters in the Mountains, etc. - Decrease of game - Blighting effects of ardent spirits upon the Indians.

There are about three hundred men, who compose the roving, hunting parties in these regions, excluding those, who remain principally at the several forts or trading posts, on the east and west sides of the mountains. One half or more of this number, are employed as "camp keepers," who perform all duties required in camp, such as cooking, dressing beaver, making leather thongs, packing, unpacking, and guarding horses, etc., and remaining constantly in camp, are ever ready to defend it from the attacks of Indians.

These men are usually hired by the company, and more or less of them accompany every party of trappers, in their excursions, or "hunts," for beaver. The trappers on the contrary, are most of the time absent from camp in quest of game, or castor.

They are divided into two classes; those engaged to the companies, to hunt for stipulated salaries; and those called "Freemen," who have horses and traps of their own, who rove at pleasure, where they please, and dispose of their furs to whom they please. They are never unhappy when they have plenty to eat. They collect skins to exchange for necessaries with the traders; their wants are few, and seldom extend beyond the possession of a few horses, traps, and a rifle, and some other little "fixens;" the attainment of these simple desires, generally constituting the height of a hunter's ambition. There are however a few individuals, who yearly lay by a small sum, from the profits of their profession, for the purpose of purchasing land, and securing to themselves a home hereafter, in some of the western states, at which they may peacefully repose in their declining years. But these instances of prudent forethought are extremely rare; and the purchase of grog and tobacco, and the practice of gaming, more frequently disperse their surplus funds, with a facility far greater than that in which they were obtained.

Many of these mountaineers have taken squaws for their wives, by whom they have children. These females are usually dressed in broad cloths, either green, scarlet, or blue. Their frocks are commonly of the latter color entirely, or a combination of the other two; the waist and sleeves being composed of one, and the skirt of the other; and these dresses appear very becoming. On their heads they wear nothing but handkerchiefs, and their feet are enveloped in moccasins. The clothing of the hunters themselves, is generally made of prepared skins, though most of them wear blanket "capotes," (overcoats,) and calico shirts. Some of them however, make coats of their buffalo robes, which are very warm and comfortable in cold weather, but become rigid and useless, if they are exposed to rains, or otherwise get wet. Moccasins are worn universally by all the whites and Indians. One half of these men are Canadians, and Half-breeds, who speak French, and some both French and English; the remainder are principally Americans, from every part of the United States. There are, however, a few foreigners, from various portions of Europe; and some Mexican Spaniards, from the Rio Del Norte; but these latter are darker, and more ill-favored than any of the mountain Indians. They are declared, by the hunters, to be an amalgamation of Indian and Negro blood, an opinion that would be pronounced by any one, unacquainted with their claim to Spanish European origin in part.

Some few of the old mountaineers annually leave the country, and their places are occupied by the less experienced "new comers," consequently the number of whites in the mountains, remains about the same.

Beaver and other kinds of game become every year more rare; and both the hunters and Indians will ultimately be compelled to herd cattle, or cultivate the earth for a livelihood; or in default of these starve. Indeed the latter deserve the ruin that threatens their offspring, for their inexcusable conduct, in sacrificing the millions of buffalo which they kill in sport, or for their skins only. The robes they obtain in the latter case, are most frequently exchanged for WHISKEY, with the traders at their establishments on the Missouri, Arkansas, and Platte rivers. - The curse of liquor has not yet visited the Indians in the mountains; but has found its way to almost all those who inhabit the plains; whose faculties are benumbed, whose energies are paralyzed, and who are rapidly sinking into insignificance and oblivion, by the living death, which their unhappy predilection for "strong water," has entailed upon them. They were gay and light hearted, but they are now moody and melancholy; they were candid and confiding, they are now jealous and sullen; they were athletic and active, they are now impotent and inert; they were just though implacable, they are now malignant and vindictive; they were honorable and dignified, they are now mean and abased; integrity and fidelity were their characteristics, now they are both dishonest and unfaithful; they were brave and courteous, they now are cowardly and abusive. They are melting away before the curse of the white man's friendship, and will soon only be known as, "The nations" that have been.

It is a prevailing opinion among the most observing and intelligent hunters, that ten years from this period, a herd of buffalo will be a rare sight, even in the vast plain between the Rocky Mountains, and the Mississippi. Though yet numerous, they have greatly decreased within the last few years. The fact is alarming and has not escaped the notice of some shrewd Indians, who however believe the evil to be unavoidable.

Chapter LX

Return home - Conclusion.

On the ninth of October we reached Belle Vue, a few miles above the mouth of the Platte river, on our return home. Here we remained eight days, in consequence of heavy rains; and in the meantime the company, consisting of about eighty men, dissolved; and each person sought such a conveyance as best suited him, to the state of Missouri. Some shipped on the Mackinaw trading boats from the upper Missouri; some made or purchased canoes, in which they embarked down the river; and others set out on foot to the completion of their journey. A small party, including myself, proceeded on horseback; leaving Belle Vue on the seventeenth. We had a cold, wet journey over a rich rolling prairie country, intersected by small streams bordered with timber, to cantonment Leavensworth, at which we arrived on the twenty-eighth. We remained here, to rest our horses and repose ourselves, two days; and on the thirty-first, after witnessing a review of the United States Dragoons here, commanded by Col. Dodge, we continued our journey. Our route lay over the same open undulating country, variagated with timbered streams; on which we observed, that as we advanced, the wooded bottoms through which they flow, increased in breadth and luxuriance.

On the twelfth of November we reached Boonville, in the state of Missouri, having been in daily view of those splendid spectacles, burning prairies, since we left the Pawnee loups. At this place I disposed of my horse, and took passage on a steamboat to St. Louis, which I reached on the fifteenth, after an absence of nearly six years.

Those who have done me the honor to peruse my journal have obtained a fair idea of the character of this eventful period of my life, and of the character of the lives of trappers in the Rocky Mountains in general. Roaming over those dark regions of solitude, constantly exposed to danger from wild animals and ravages; frequently obliged to endure the most severe and protracted privation and fatigue; separated by many hundred weary miles from the abodes of civilization and refinement; conscious that even the fond filaments of love and consanguinity, which wind in delicate fibres around the heart, were becoming attenuated by distance, time, and novelty; feeling that my habits and manners were gradually giving way to the innovation of savage and unsocial custom, and my very speech and person were yielding slowly, but not less surely, before the uncouth barbarisms of language, and the exposures and severity of an ever-changing climate - it will not seem strange that I sometimes repined at my long absence from the scenes of my nativity, and reviewed with regret the inducements that forced me from friends and home. But these, however, were not always my feelings; - resolute, cheerful, contented, I usually was. And when the weather was warm and pleasant; the demands of nature satisfied, a reliance on the good qualities of my arms and ammunition, not misplaced; the confidence of bestriding and governing a truly noble steed, in the spirit of stirring excitement of the chase, gloriously bounding over the plains, in the panoply of speed and power, before which the swiftest and mightiest denizens of the forest and prairie must yield themselves victims; then - then I was really, rationally happy. Many times have I experienced the sensations, generated by either condition; but these scenes have now passed away, their delights and perils no longer thrill nor alarm, and I bid them farewell forever.

Chapter LXI

The Rocky Mountain Indians are generally of the middle stature, straight, well proportioned, have fine limbs, black hair flowing loosely over their shoulders, in some instances so long as to reach the ground when they stand erect, and lively keen black eyes. Their features are seldom ugly, often agreeable, and in smiling they exhibit beautiful, white even, set teeth; except those who subsist principally upon fish. Their clothing consists of a long shirt, reaching down to the knees, open on each side from the armpits to the bottom; a pair of legging a breech-cloth and moccasins, over all of which a light buffalo robe or blanket is thrown, resting on the shoulders. - The shirt and leggins are usually made of the skins of Rocky Mountain sheep, dressed for that purpose with peculiar care; and are ornamented with small blue and white beads, colored porcupine quills, and leather fringe over the seams, or instead of the latter, human hair dyed of various hues, which is obtained from the scalps of their enemies. They paint their faces with vermillion, white earth, ochre, powder, etc., and fasten beads, shells, buttons, and other trinkets, with feathers in their hair. The garments of the women, are a long gown, also made of the smooth even skin of the Big-horns, decorated about the neck and shoulders with all the beads they can procure, short leggins, moccasins, and a light robe thrown loosely, yet gracefully, over their shoulders. Their dresses are sometimes loaded with some eight or ten pounds of large cut glass beads of various colors; their leggins and the tops of their moccasins are also ornamented with porcupine quills, or small beads. Like the men they are very fond of decorations, vermillion, trinkets, shells, etc., but they attach them to their dresses, instead of their hair; and like them besmear their faces with the former.
Nothing can divert an Indian from his purpose, when in pursuit of game; but on

his return to camp he sinks in comparative indolence. They are brave, and display both intelligence and deep artifice, in their stratagems to surprise and conquer their enemies; but mercy, the hero's noblest attribute, is to them unknown, at least unregarded. Sedate and taciturn, with folded arms, they slowly pace back and forth before their lodges, to all appearance, wrapt in intense thought; approach a warrior thus intellectually engaged, and inquire the subject of his meditations, he will answer with undiminished gravity, "nothing." In their lodges and domestic circles, they are loquacious as the whites, with the difference of good-breeding, that the person speaking is never interrupted; and in turn each speaks, and is listened to with profound attention. Their conversations most frequently are on themes of war, and each individual chaunts his own exploits. Universally superstitious, they put good or ill constructions on their dreams, and account for all the phenomena of nature, by attributing all effects, to the kind or unfavorable temper of the Deity. With a commendable devotion they believe unhesitatingly all the Great-Spirit thinks proper to communicate to them, through their "Medicine men," (i.e. priests,) who, taking advantage of their religious credulity, impose upon them the most ridiculous stories of divine truth.

These medicine men also forewarn them of coming events, and profess to cure them of their diseases, by wearing the skin, assuming the characters, imitating the voices, and mimicking the actions of bears, and other animals; accompanying their demoniac capers by discordant yells, and deafening sounds, extracted from a kind of drum by violent thumpings, alone sufficient, one would imagine, to frighten away both the demon of distemper, and the spirit of the afflicted together. But their endeavors are probably well meant; and they may, from the habit of relating their powers, perhaps have induced themselves to believe they really possess the qualifications, that they arrogate to themselves, and which are so readily conceded them, by their less gifted comrades, - at least they may claim the merit of performing, without any fees, all the various ceremonies of their profession; their only reward being the esteem of their companions, and a more extended influence. Their diseases are however few, and when slightly ill, they usually resort to their sweat houses; which are constructed of willows, pointed and stuck in the ground, with their tops bent over and interlocked, so as to form a hemisphere of about six feet in diameter, on which skins, blankets, or robes, are spread to prevent the escape of the steam. In the centre of this hut a small hole is excavated in the earth, over which dry sticks and stones are placed in alternate layers, and fired beneath; when the wood consumes, the heated stones fall to the bottom, and the invalid, accompanied by others who do so for amusement, divested of all clothing, enter the cabin with a basin of water, which they commence pouring slowly on the stones, and soon fill it with dense scalding vapor; when it becomes no longer supportable they cease pouring, but recommence when it subsides. By this means they produce a copious perspiration from their bodies, which continues until they leave the hut, in which they sometimes remain half an hour. - When they crawl out, their bodies appear half boiled, and they immediately jump into the nearest stream, remaining immersed in the water until they become chilled in as great a degree as they were before heated. Indeed I have often seen them bathe in the winter season, through a hole cut in the ice, after issuing from a sweat house. - This practice, though it may bring present relief, must be attended in the end, with highly deleterious consequences; and indeed I have often observed that young Indians, who were in the habit of taking this hot and cold bath weekly, have foreheads wrinkled, and many other indications of premature old age. Its effects are more apparent of course, in those of mature years. Bathing is one of their favorite amusements, and when near a suitable place, if the weather be fair, some of them may at any time be seen in the water. The employments of the men are war, hunting, fishing, guarding their horses, shooting at a target, horse racing, etc.

Chapter LXII

The women of most tribes in the Rocky Mountains, though sometimes pretty when young, soon become decrepid and ugly, in consequence of the great hardships they are compelled to endure throughout their whole lives. They are esteemed rather as beasts of burden than companions for their tyrannical husbands; and are seldom valued as high as a favorite horse, for which they have been exchanged. They perform all the camp labors, such as drying meat, dressing skins, collecting fuel, bringing water, cooking, making clothing and moccasins, packing and driving horses from one encampment to another; and frequently accompany their husbands in the chase, to butcher and bring in the game they may happen to kill. Notwithstanding the constant drudgery to which they are subjected, they appear cheerful, sing during the performance of their duties, in rather a wild mournful air, and seem contented to wear through life, patient as an ox, submitting to all the hardships their capricious lords impose upon them without a complaint, which, however, instead of exciting sympathy or pity, would only exact blows and abuse from them.
Infants, when they first appear on life's eventful stage, are immediately immersed in a snow-bank, in a state of perfect nudity, a few moments, for the purpose of familiarizing them to the endurance of cold; if in the summer, they are instantly plunged into water, perhaps to render them as nearly as nature will permit, amphibious: if it be a female child, it is regarded by the father with contempt, and valued little more than a puppy; on the contrary, if a male, it is caressed and fostered by him, and the utmost pains taken to instil into its mind, as soon as it becomes able to comprehend, a love of glory, spirit of chivalry, and a contempt of privation, and hardship, and suffering, to which he must necessarily be exposed in leading the life of a hunter and a warrior. He is taught by example, the precepts early inculcated, and daily practised by his father, but is on no occasion punished, or deprived of any gratification. His youth is passed in hunting small animals and birds, in gymnastic exercises, such as are calculated to render him robust, active, and fearless, and in riding on horseback, with various

equestrian feats, an accomplishment essential to his future character and destiny. When he becomes old enough to enter the lists of battle, he assembles, with other youthful aspirants for glory, chaunts the inspiring war song, and proudly dances in the midst of his companions, vowing vengeance against the enemies of his race; and shouting to the charge, enacts what his enthusiastic imagination prescribes his duty on the field of honor, and yells of rage and anger fill the air, as bursting from his lips in fiery wrath. The spectators applaud, and hardy veterans, kindling into zeal, join in the dance, proclaim aloud their achievements, exhibit their trophies, brandish their weapons, and with vehement eloquence encourage the high-souled youth to follow their footsteps, smite his enemies with vindictive fury, and return triumphant to his tribe and kindred, by whom he will be welcomed with shouts of rapture and applause, and win the favor of Heaven. His bosom glows with exultation, and burns with ardor to emulate the glowing deeds of his ancestors; to win for himself an honor and a name wholly, peculiarly his own, and in addition, secure to himself a happy enhance to the land of spirits, all of which he believes are to be won on the field of battle. Thus in his fervor, happiness, renown, immortality, are believed to be the reward of valor, and an impression is formed that will last with his life; he resolves to be a brave, and becomes - nay is - a warrior.

War dances are usually held in the evening, round a large fire, made for the purpose; and though the performers are entirely divested of clothing, they are attended by women and children; accustomed, however, from infancy to exhibitions of this kind, they are unconscious of there being any impropriety or indecency in them. A large circle being formed by the bystanders around the fire, the ceremony commences with a general song, in which they all join in perfect harmony, beginning in the highest strain their voices are capable of, and gradually falling note after note, until the sound dies imperceptibly away; then suddenly resuming a high shrill tone, and again falling gradually as before, the song being accompanied by a regular thumping on a discordant kind of drum. In the mean time the naked youths and warriors jump into the ring, and dance with vigor, keeping time with the drum, yelling at intervals like demons, shaking aloft their gory gleanings of a field of carnage, (scalps, etc.) flourishing their weapons in their rapid evolutions, throwing their bodies into unnatural, sometimes ludicrous attitudes, and chaunting aloud their deeds of desperate daring. These dances are invariably celebrated on the return of a successful war party, and at such other times as their inclinations may direct.

When a young man wishes to marry, he selects a young woman, and makes application to her parents, who, if they think he can support a wife, will probably demand a horse, which if he promises to bestow, concludes the contract, and he is invited to come and sleep with her, at her father's lodge, without perhaps having ever had any conversation or acquaintance with her. In such cases, the wishes, however, of the young woman are always ascertained by her mother, but if averse to the proposed bride-groom, not always followed. Sometimes he sends a present of one or more horses to the favored damsel's parents, and if they are accepted, he sends a horse to her lodge the first time they move camp, which she sides; from that moment she is considered his wife, and at night encamps with him.

Polygamy, though allowed, is by no means common among them, but in accordance with the customs of their forefathers, they put away their wives and take others at pleasure. I have known an Indian to have fourteen wives, all living; but this is a solitary instance, and similar ones occur very rarely. In most cases, the husband and wife pass their lives agreeably, and sometimes happily together, and when separations take place, they proceed most commonly from the infidelity of the latter not the inconstancy of the former. The females thus banished from the lodges of their husbands, return to their parents, by whom they are received and supported until another opportunity occurs to marry.

It is the custom in some nations, on the death of a friend, to cut off a joint of a finger; and in others, to clip off their hair and continue mourning until it grows out again to its usual length; others, again, black their faces with a composition that only wears off with the skin, and cease to mourn when it disappears. But all nations give utterance to their grief in a piercing, wild, monotonous lamentation, commencing at a high key and gradually falling to the lowest notes, in continued repetition, that sounds extremely melancholy, and fills the hearer with solemnity.

They all inter their dead in the earth, with the exception of infants, who are sometimes encased in skins and deposited in trees; but are generally buried like adults in the ground.

Chapter LXIII

Of their vague and crude notions of the past and of futurity, each tribe in the Rocky Mountains has its own favorite traditions, most of them too ridiculous either to hear or repeat. Some believe that a beautiful woman descended from the clouds, at an era when Indians were but few, and even those extremely poor; that she supplied the country, before destitute, with game, and promised that brave men and warriors, who perished on the field of battle, should go to a pleasant country after death, where there is very little snow, good horses, and plenty of game, not very wild; she then disappeared. Another tradition speaks of a wonderful beaver that suddenly appeared in a certain river, of immense size and great sagacity, gifted with intelligence and speech, which bestowed great bounties on the people, promised them happiness in a fine country after death, where game was abundant and hunting easy; when he also vanished. A third informs us that a prairie wolf visited the sources of the Columbia, ages ago, and finding the people poor and miserable, caused the mountains to be stocked with deer, sheep, goats, elk, moose, etc., and the plains with antelopes, and the rivers with salmon; so that the people, who had before lived wretchedly upon roots and berries, were overjoyed, and loved the wolf; which afterwards married a cow, and had innumerable difficulties with a rival bull, which he finally succeeded in destroying, and lived long afterward happy and undisturbed; but at length, he promised the people happiness in a fine hunting country after death, where brave men would be loved, and have much less trouble in their excursions for game than cowards, who would always be miserable and distressed. With these words he left them, and was seen no more. These, and many other equally absurd traditions, are the grave ideas of the mountain savages on this subject. All agree, however, in the belief of a future state of existence, more happy to the brave, than the present; and in a great spirit, by whose bounty they will be permitted to inhabit that delightful region.

Such of the Indians as possess horses enough to convey themselves, their families, and their baggage, with ease, are esteemed wealthy; their animals are guarded with peculiar care, during the day in bands of twenty or more together, by boys; and in the night, they are tied to stakes, driven in the ground for that purpose, near their lodges, for greater security against robbers. Those who are not so fortunate or wealthy as to possess the number of horses requisite, are obliged to walk or put enormous loads upon such as they may chance to own. In one instance, in the year 1832, I saw a mare loaded with, first - two large bales containing meat, skins, etc., on opposite sides of the animal, attached securely to the saddle by strong cords; secondly - a lodge, with the necessary poles dragging on each side of her; thirdly - a kettle, axe, and sundry other articles of domestic economy; fourthly - a colt too young to bear the fatigue of travelling was lashed on one side; and finally - this enormous load was surmounted by a woman with a young child; making in all, sufficient to have fully loaded three horses, in the ordinary manner. Though this rather exceeds any thing of the kind I ever before saw, yet large loads, in like manner surmounted by women, children, colts and puppies, are often observed in their moving jaunts.

Some of the poorer classes, who do not possess horses, and are consequently unable to follow the buffalo in the prairies, ascend the mountains where deer and sheep are numerous, and pass their lives in single families - are never visited by the horsemen of the plains, but sometimes descend to them, and exchange the skins of those animals for robes, and other articles of use and ornament.

Their weapons are bows, arrows, and war-clubs, and are of their own manufacture. Their bows never exceed two and a half feet in length, and are made usually of the rib bones of the buffalo, two of which, in the construction of this weapon, are neatly jointed, glued together, and wound with thongs about the joint; it is gradually tapered from the middle toward each end, is polished, and rendered more elastic by sinews glued on the back, from end to end, over which rattle snake skins are sometimes cemented for ornament. The string is always composed of sinews, twisted together into a cord. Bows are made sometimes of elk horn, and sometimes of wood, but are always strengthened by adding sinews to the back, and not, as an eminent western writer has observed, "by adding buffalo bones to the tough wood."

Their arrows (except the poisoned ones of the Sann pitches) are made of wood, slender, never above two feet in length, and are pointed with sharp transparent flints, neatly broken to a dagger-blade shape, from half to three-fourths of an inch in length, which never exceed the latter. These points are ingeniously inserted in a slit in the end of the arrow, are fastened by sinews wrapped around it, and are rendered less liable to damage by being covered with a coat of glue. They have three or four distinct feathers, six or seven inches in length, placed opposite to each other, remaining parallel, but turning gently on the arrow, in order to give it a spiral motion, which prevents its wavering, and enables it to cleave the air with less resistance.

They manufacture spears and hooks, also of bone, for fishing, but they are not to be compared to the same instruments made of metal by the whites. But they have been supplied by the traders with light guns, spears, and iron arrow points, which have in measure superseded their own weapons; still, however, bows and arrows are most frequently employed in killing buffalo.

Chapter LXIV

The Indians rise at day light invariably, when all go down to the stream they may happen to be encamped on, wash their hands and face, and comb their long hair with their fingers, the nails of which are allowed to grow long.

The men are all expert horsemen, ride without saddles generally, always when they pursue game. Their saddles are made of wood, sometimes of elk horn and wood, covered with raw hide, which renders them very durable and strong; their stirrups are also composed of wood. - Instead of a bridle and bit, they make use of a long cord, (used for catching wild horses) tied to the under jaw of the animal; these cords, or more properly leashes, are made of narrow strips of raw hide, divested of the hair, and rendered soft and pliant by greasing and repeated rubbing; or of the long hair from the scalp of the bison, twisted into a firm and neat rope. The saddles of the women have both before and behind, a high pommel, reaching as high as their waists, for the greater convenience of carrying children. Some of them carry two children and drive their pack horses. They drop their little ones on the ground, dismount and arrange their packs, when necessary, remount, ride along side of them, and draw them up again by the hand, and continue their march. Their children of three or four years of age, are lashed firmly on the top of their packs, and are often endangered by the horses running away with them, though I never saw one seriously injured in consequence. Yet, if the loads turn under the horses' belly, as they sometimes do, the situation of the poor child is truly frightful.

Gaming is common with all the mountain Indians. Some of them, however, are more passionately addicted to it than others. The games mentioned in my journal, and described as having been observed among the Flat-heads, are those practised by all the various tribes, without exception.

Indian sagacity has ever been a riddle to the greatest portion of the public; who have been inclined, in some degree, to consider their great skill in tracing an enemy, or detecting the vicinity of danger, as depending on a finer construction and adaptation of some of the organs of sense, and to believe that no white man could ever attain a degree of skill by any means equal to the tact so often exhibited by the aborigines of this country. But I am convinced it is entirely owing to a careful attention to, and observation of, various minute peculiarities, which would escape the notice of any one not instructed how to prosecute his examination, though ever so attentive; and in proof of this, there are in the mountains many hunters who have become so practiced in this species of discernment, as to be acknowledged by the Indians as equal to themselves; though these Indians, in this respect, are inferior to none.

Several tribes of mountain Indians, it will be observed, have names that would be supposed descriptive of some national peculiarity. Among these are the Black-feet, Flat-heads, Bored-noses, Ear-bobs, Big-belly's, etc., and yet it is a fact, that of these, the first have the whitest feet; there is not among the next a deformed head; and if the practice of compressing the skull so as to make it grow in a peculiar shape ever did exist among them, it must have been many years since, for there is not one living proof to be found of any such custom. There is not among the Nez-perces an individual having any part of the nose perforated; nor do any of the Pen-d'orulles wear ornaments in their ears; and finally, the Gros-vents are as slim as any other Indians, and corpulency among them as rare.

The funerals of the mountain Indians are similar to those described by various authors of other nations. The dead body is dressed in the neatest manner, envelloped in skins, and deposited in the earth, with the weapons, ornaments, trinkets, etc., which belonged to the deceased when living; and it is customary among some tribes, to sacrifice his favorite horse on the grave of a warrior, after the manner of the Tartars.

Chapter LXV

The Crow Indians call themselves "Ap-sa-ro-ke," and are tall, active, intelligent, brave, haughty and hospitable. They acknowledge no equals, and pride themselves on their superiority over other nations. They boast of greater ability in making their wishes understood by signs, to those who do not comprehend their language. - In their lodges, they consider themselves bound to protect strangers, and feast them with the best they can procure. They are well armed with guns and ammunitions, which they have purchased from the whites, or taken from their enemies; and are supplied with horses, which they steal from other nations, with whom they are constantly at war.

Their women are, without exception, unfaithful, and offer themselves to strangers unhesitatingly for a few beads or other trifles. Jealousy is hardly known among them, though the property of one proved to have had illicit connexion with another's wife, by "crowic law," is immediately transferred to the injured husband.

War parties, on foot, are frequently sent out, at all seasons of the year, for scalps and property; and every man in the village is compelled, in turn, to join these predatory excursions. They are slow to project, cautious to proceed, certain to surprise, firm to execute; and are ever victorious when they commence an affray. Extremely superstitious, with certain charms about their persons, such as the skins of birds, small animals, or other trifles, in the efficacy of which they place great reliance, they charge fearlessly upon their enemies, under the entire conviction that these their medicines ward off every danger, and completely shield them from harm.

They are good hunters and accomplished horsemen; they ride at full speed, throw their bodies entirely on one side of their animals, shoot from the same side under their horses' necks, and pick up their arrows from the ground without abating their rapidity.

They are fond, nay, proud of ornaments for their persons, particularly scarlet cloth, feathers, and vermillion, with which they attire themselves with great care for the field of battle; but they are fonder, prouder still, of a high reputation for bravery and prowess.

Among the trophies of victory, a scalp is prominent, and belongs, not to him who shot down the enemy, but to him who first rushed upon him and struck him on the head. The very pinnacle of human glory with them, is to have taken twenty scalps from their enemies, and to achieve this high honor is the constant ambition of every noble-minded brave.

They express their gradations of rank by affixing ornaments to their hair, among which the tail feathers of the golden or calumet eagle, indicate the loftiest. None but the most heroic warriors are entitled to the envied privilege of wearing this valuable plumage, (a single feather of which is more prized than a fine horse) and the person in whose hair even a single quill is conspicuous, is entitled to, and receives, the most profound respect and deference; but when a plurality of these badges are displayed by some fortunate brave, he is regarded with veneration, almost with awe.

They are entitled to precedence in the science of theft, of which they are masters; being almost certain to obtain from a stranger any article they discover about him, that they may chance to desire. When detected in pilfering, they appear extremely foolish and ashamed, because it is superlatively dishonorable - not to steal - but to have failed in an attempt, the complete achievement of which would have added renown to their characters for dexterity and address.

They are now the declared enemies to trappers, several of whom they have killed; yet there is an establishment on the Big-horn river belonging to the American Fur Company, at which they trade peaceably. During the summer of 1835, they caught two of our trappers, at the point of the Wind Mountain, detained them several days, robbed them, and then allowed them to return to us. It must be regarded, however, as a freak of kindness, not as a settled principle of mercy. They rove in two villages, containing six or seven hundred souls in each. These villages seldom leave the Yellow Stone and its tributaries, but war parties infest the countries of the Eutaws, Snakes, Arrapahoes, Blackfeet, Sioux, and

Chayennes, with all of whom they are at war. They have a tradition that their forefathers fought a great battle with the Snakes, in which many hundred of the latter were killed; that the battle lasted three days, during which time the Chief of the Crows caused the sun to stand still.

Several years since, one of the Crow villages, issuing from a narrow defile into a large prairie valley, through which they designed to pass, discovered a village of Blackfeet encamped on the opposite side. A council was immediately held, in which it was resolved to place a few lodges in a conspicuous situation, purposely to be observed, and conceal the greater part, together with all their war horses, in a low place where they could not be seen. This was done, and their inferior horses were permitted to stray down the plain, offering an inducement to their foes to come and take them. At the same time, the Blackfeet discovered, as they supposed, a few lodges of the Crows, and thinking themselves undiscovered, made instant preparations for combat. These young warriors were sent in a body to surround the Crows in the night, and directed to commence the attack at daybreak, at which time the old braves were to join them on horseback, and assist in exterminating the enemy. In obedience to their instructions, the young men set out that evening, but the night being dark, they mistook their way, and were prevented from reaching the field of glory at all. - Early in the morning, the old veterans, believing that the business would be half finished before they could reach the scene of action, set off at full speed, which they continued unbroken until they reached the fatal spot. The Crows allowed them to approach to the very doors of their lodges, when the war-whoop resounded in their ears, and in a moment every hill was red with mounted warriors, bearing down upon them. They saw the snare into which they had fallen, when too late, and fled; but alas! their horses, already fatigued, were unable to run. The Crows pursued them on fresh horses, overtook and killed them as they fled, until the survivors gained the mountains. - Seventy scalps were displayed on that day by the victors, who also obtained considerable booty. So much for out-manoeuvering and discomfiting their enemies.

The following anecdote exhibits their character in a different light. During a long continued storm in the winter of 1830, the Sioux Indians carried off nearly one hundred and fifty head of horses from them, and a party of eighty Crows immediately departed in pursuit. They travelled several days in the direction that they supposed the Sioux had taken; but the protracted storm prevented them from discovering the traces of their enemies, and all but twenty-three returned home. The small party still in pursuit, consisted of men who scorned to return without even a glimpse of their enemies, and continued on, far beyond the distance that horses could have travelled at that season of the year, in the same length of time. Finally concluding that they had quite mistaken the course of their enemies, they resolved to retrace their steps towards their own firesides; but some days after, unfortunately, they passed, during a tedious snow storm, very near a Sioux village, and were discovered by them. A large party were soon in pursuit, and the Crows, unconscious of danger, advanced at their usual pace until night, and encamped as customary. During the night, they were entirely surrounded by the Sioux, who awaited for the next day to make an attack upon them. In the morning they made the necessary preparations for departure; breakfasted, packed their dogs, and started; but the Sioux, till this moment unobserved, rushed towards them, and drove them into a hole, which had been the bed of a spring torrent, crossed by a fallen tree. To this they suspended their robes for concealment. In the mean time, a Sioux chief, who could speak the Crow language, stepped in front of the rest, and in a loud voice addressed them, saying that he was very sorry to keep them in suspense, but that his party were nearly frozen, having passed a sleepless night in the snow without fire; that they were consequently compelled to kindle a fire and warm themselves before they could proceed; "but," continued he, "as soon as we thaw ourselves, you will cease to exist, unless you are what your name indicates, and fly away; or like the prairie dog, you can burrow in the ground." - Accordingly, they warmed themselves, attacked the devoted little band, and killed them all, with the loss of but two men killed, and several slightly wounded. Three of the Crows fought manfully, but the others, benumbed as they were by the cold, could make but little resistance. Their bodies were cut in pieces, and transported to the village on dogs, the same day.

Chapter LXVI

The Snake Indians are termed in their own dialect, Sho-sho-ne. They are brave, robust, active and shrewd, but suspicious, treacherous, jealous and malicious. Like the Crows, war-parties frequently go out to plunder, and some times kill and rob the trappers if they think it will not be discovered. They are utterly faithless even to each other.
There is one evil genius among them, called the "Bad Gocha," (mauvais gauche - bad left-handed one) who fell in with a party of trappers, led by a well-known mountaineer, Mr. E. Proveau, on a stream flowing into the Big Lake that now bears his name, several years since. He invited the whites to smoke the calumet of peace with him, but insisted that it was contrary to his medicine to have any metallic near while smoking. Proveau, knowing the superstitious whims of the Indians, did not hesitate to set aside his arms, and allow his men to follow his example; they then formed a circle by sitting indiscriminately in a ring, and commenced the ceremony; during which, at a preconcerted signal, the Indians fell upon them, and commenced the work of slaughter with their knives, which they had concealed under their robes and blankets. Proveau, a very athletic man, with difficulty extricated himself from them, and with three or four others, alike fortunate, succeeded in making his escape; the remainder of the party of fifteen were all massacred. - Notwithstanding this infernal act, its savage author has been several times in the camp of the whites; but his face not being recognized, he has thus far escaped the death his treacherous murder so richly merits. As the whites are determined, however, to revenge their murdered comrades, the first

opportunity, there is little doubt that his forfeit life will, sooner or later, pay the penalty of his crimes to injured justice! During the life of that white man's friend, the Horned Chief, Gocha was seldom if ever seen with the Snake village, but was heard of frequently, with a small band of followers, prowling about the Big Lake. The principal chief of the Snakes is called the "Iron Wristband," a deceitful fellow, who pretends to be a great friend of the whites, and promises to punish his followers for killing them or stealing their horses. The "Little Chief;" a brave young warrior, is the most noble and honorable character among them. Notwithstanding the bad qualities of these Indians, their country is rich in game, and the whites have thought proper to overlook many serious offenses, rather than expose small trapping parties to the vindictive attacks that would characterize an open war.

It is hardly necessary to say, that though not so completely adept in the art of stealing as the Crows, they are nevertheless tolerably expert. I saw one of them steal a knife without stooping from his upright position, or in the least changing countenance. He was barefooted, and connived to get the handle of the instrument between his toes, and then draw his foot up under his robe, where a hand was ready to receive the booty.

There is one Snake who has distinguished himself in battle, called "Cut Nose," from a wound he received in the facial extremity, from the Blackfeet, who has joined the whites, lives with and dresses like them. He is an excellent hunter, and has frequently rendered them important services.

The females are generally chaste, the men extremely jealous. I have heard of but one instance of adultery among them, which was punished by the enraged husband by the death of the female and her seducer, the latter of whom was murdered, and then placed upright on a high cliff, in a valley on Horn's fork, as a warning to all, and a fearful monument of an injured husband's revenge.

Of the Snakes on the plains, there are probably about four hundred lodges, six hundred warriors, and eighteen hundred souls. They range in the plains of Green river as far as the Eut mountains; Southward from the source to the outlet of Bear river, of the Big Lake; thence to the mouth of Porto-muf, on Snake river of the Columbia; and they sometimes ascend the Snake river of the Soos-so-dee, or Green river, and visit the Arrappahoes, on the sources of the Platte and Arkansas. They are at war with the Eutaws, Crows, and Blackfeet, but rob and steal from all their neighbors, and any body else whenever an opportunity occurs. It may be said of them that stealing is their master passion; and indeed they are so incorrigibly addicted to the practice, that they have been known in some instances to steal, in the most adroit manner imaginable, even their own horses, mistaking them, of course, for the property of others.

Chapter LXVII

The Eutaws are neither ugly nor prepossessing. They are brave, but extremely suspicious; are candid, never treacherous, and are not inhospitable, though they rarely feast a stranger who visits them, like the Crows, Arrappahoes, or Blackfeet, but in general ask for something to eat themselves of their guests. Unlike the Crows or Snakes, they never steal; but they are most accomplished beggars. With unblushing assurance, one of them one day asked me for my coat, and was of course refused. "Then give me your blanket" - denied; "your pistols, then," - no; and was going to ask for every article he saw about me, but ascertaining by a very abrupt reply, that I was not inclined to listen to him, and still less to give him any thing, he turned to one of his companions and observed, "that man's heart is very small."

The reader no doubt remembers the ingenious impudence of La Fogue in directing his children to weep, and then telling us they were crying for something to eat, with a pathetic gravity, to obtain a little food, when there were in his lodge at the time several bales of dried buffalo meat, and we had nothing else, he well knew. It is not, however, so much for the property itself that they beg with incorrigible pertinacity on every occasion, as they do invariably; but for the honor of having obtained it in that mode; for to be adroit and successful in the art is considered a demonstration incontestible of uncommon talent.

They are very jealous, though their women give them little cause for being so. They are as well clad, but are not so well supplied with lodges as some other Indians, because there are no buffalo in their country, and they are obliged in winter season to construct cabins of cedar branches, which are by no means so comfortable.

They are, by far, the most expert horsemen in the mountains, and course down their steep sides in pursuit of deer and elk at full speed, over places where a white man would dismount and lead his horse.

They are less addicted to gambling than most other Indians; probably, because they subsist principally upon small game, more difficult to obtain, and hence have less leisure for other pursuits. In places where deer are numerous, they excavate holes in the earth, in which they conceal themselves, and shoot them as they pass in the night.

They frequently visit Taos on the Del Norte, and California; consequently many speak, and nearly all understand, the Spanish language.

There are of the Eutaws, in all, about two thousand five hundred souls.

The Arrappahoes

are a tribe of Indians who rove in one or two villages, as their inclinations may indicate, on the sources of the Platte and Arkansas rivers. They were for several years deadly enemies to the whites, but Capt. Ghant, whose firmness and liberality they have reason to remember long, has established a trading house among them on the Arkansas, four day's march from Taos, and has succeeded in gaining their confidence and inculcating among them, an opinion of the whites, that will perhaps secure their lasting friendship. They set before their guests the best provisions they can procure, and feel bound to protect them even at the risk of their own lives. They consider hospitality a virtue second only to valor.

Polygamy is common among them, every man having as many wives as he pleases.

The flesh of dogs is esteemed with them a dainty, and by some of the neighboring nations they are detested for no other reason, than because they eat them.

They are brave, candid, and honest, for though good warriors, they neither rob nor steal. Some of Mr. Vanderburgh's men wintered with them in 1834 and '35, and speak highly of their integrity and generosity. They lost nothing by theft during their stay, but were kindly treated, and the chief told them on separating in the spring, to display a white flag on a pole near their encampment, and they would never be molested by the Arrappahoes. As their country abounds with buffalo, they have fine lodges and live comparatively easy and comfortable. - Their number is in all about two thousand souls, and they are friendly with the Snakes, Black-feet, Gros-vents, and Comanchies; but at war with the Eutaws, Crows, and Sioux.

Chapter LXVIII

The Grizzly Bear

This animal, well known to be the most formidable and dangerous in North America, is found in the Rocky Mountains of every shade and color, from black to white, but most frequently of a dark grizzly hue, and is so large that those of a brown color are often mistaken by hunters who are accustomed to see them almost daily, for buffalo.

For instance - returning from the waters of the Columbia to those of the Colorado, in the spring of 1834, with four white men and several Indians, on our way over Horse Prairie, I discovered and passed within one hundred yards of a very large female, accompanied by two cubs, but was prevented from firing upon her from the fear of arousing some lurking foe in this dangerous part of the country, and permitted her to pass without molestation. At our encampment in the evening, the men, who usually discuss the occurrences of the day over their suppers, surprised me by inquiring if I had ever before seen a female bison with two calves. I replied that I had not yet met with an instance of the kind. "What?" exclaimed all of them, "did you not see a cow followed by two calves, who passed us this afternoon at ____ ?" describing the place - when I immediately discovered that they had mistaken the bear for a buffalo, and found them still incredulous when I asserted the fact.

A trapper, with whom I am acquainted, discovered, as he supposed, several years since, a herd of buffalo near Bear river, and being without provisions, resolved to approach them. Accordingly he followed the winding of a deep gully, which at length brought him among them, but imagine his disappointment and terror, when in lieu of buffalo, he found himself surrounded by thirty grizzly bears, whose aspects were any thing but amiable. Perceiving, however, that he was yet

undiscovered, he retraced his steps, and fortunately escaped a horrible death, which in all probability he would have suffered had they observed him.
I have often mistaken them for buffalo, and discovered my error only when they erected themselves to ascertain what passed near them, which they always do when they hear, see, or smell any thing unusual.
They are numerous throughout the mountains, particularly when fruit is most abundant, which serves them for food in the fall; roots being their chief subsistence in the spring. Through-out the long inclement winter season, like the black bear, but unlike the white, they penetrate and lay dormant in caverns. Most animals, from their superior speed, can escape them; hence, though extremely fond of flesh, they kill very few; yet they often rob wolves of their prey, and devour it at leisure; and sometimes, but seldom, catch a deer or elk that happens to pass near without discovering them.
When wounded they are terrible and most dangerous foes; and unlike all other animals, are so extremely sagacious and vindictive, that though their enemy is concealed from their view, they will rush to the spot whence the smoke of his gun arises in search of him, and if he has not already secured his safety, he will hardly have an opportunity.
Many of the trappers bear the most incontestible proofs of having been roughly handled by them, of which the most shocking instance, together with attendant circumstances, is told of a well known trapper by the name of Hugh Glass, which is so extraordinary, that I shall give a brief sketch of it here. The same incidents have already been related, I believe, in the Southern Literary Messenger.
Glass was an engaged trapper in the service of Major Henry, who was the leader of a party of beaver hunters several years in the mountains. In the year 1822 or '23, during an excursion to the sources of the Yellow Stone, Glass was employed in hunting for the subsistence of the company. One day, being as usual in advance of his friends, in quest of game, he reached a thicket on the margin of a stream, which he penetrated, intending to cross the river, as it intersected his course. But no sooner had he gained the center of the almost impenetrable underbrush, than a female bear, accompanied by her two cubs, fell upon him, cast him to the ground, and deliberately commenced devouring him. But the company happening to arrive at this critical moment, immediately destroyed the grizzly monsters, and rescued him from present death, though he had received several dangerous wounds, his whole body being bruised and mangled, and he lay weltering in his own blood, in the most excruciating agony. To procure surgical aid, or to remove the unfortunate sufferer, were equally impossible; neither could the commander think of frustrating the object of his enterprise by remaining idle, with a large party of men, engaged at high salaries. Under these circumstances, by offering a large reward, he induced two men to remain with, and administer to the wants of poor Glass, until he should die, as no one thought his recovery possible, and proceeded on with his party to accomplish the purpose of the expedition. These men remained with Glass five days, but as he did not die perhaps as they anticipated he soon must, when the company left them, they cruelly abandoned him, taking his rifle, shot-pouch, etc., with them,

believing that he would soon linger out a miserable existence. Leaving him without the means of making a fire, or procuring food, the heartless wretches followed the trail of the company, reached their companions, and circulated the report that Glass had died, and that they had buried him. No one doubted the truth of their statement until some months afterward, when to the astonishment of all, Glass appeared in health and vigor before them; but fortunately for one of the villains, he had already descended the Missouri and enlisted in the service of the United States. The other, though present when Glass arrived, being a youth, received a severe reprimand only from the justly exasperated hunter, for his unpardonable crime.

After Glass was deserted, he contrived to crawl, with inexpressible anguish, a few paces to a spring, the waters of which quenched his feverish thirst, and a few overhanging bushes loaded with buffalo berries and cherries, supplied him with food for ten days. Acquiring by degrees a little strength, he set out for Fort Kiawa, a trading establishment on the Missouri, three hundred miles distant, a journey that would have appalled a healthy and hardy hunter, destitute as he was of arms and ammunition. By crawling and hobbling along short distances, resting, and resuming his march, and sustaining life with berries and the flesh of a calf, which he captured from a pack of wolves and devoured raw during his progress, he finally reached the fort and recovered his health. After innumerable other difficulties and adventures, Glass finally fell a victim to the bloody-thirsty savages on the Missouri.

There is also a story current among the hunters of two men, whose names I have forgotten, who being one day some distance from the camp, discovered a large bear, busily engaged in tearing up the ground in search of roots. One of them, who was reputed an excellent marksman, desired his companion to ascend a tree, and after selecting one as a refuge for himself in case he should only wound and enrage the bear, deliberately elevated his piece and fired. In an instant the huge and vindictive beast rushed like a tiger toward him, seized him in the act of ascending the tree, a branch of which had caught his coat, and prevented him from accomplishing his purpose, and tore him in pieces. Having thus consummated his vengeance, the bear sank down beside his victim and expired. The other person, terrified at the bloody scene enacted beneath him, descended and ran to camp, informed his comrades of the melancholy circumstance, and returned to the spot with them. They interred the remains of the ill-fated hunter, and on examining the bear, found that he was shot directly through the heart. Hunters usually shoot them through the head, when, if the ball is well directed, they always expire instantly; yet they dare not molest them unless they have the means of escaping, either on their horses or by ascending trees. I have often heard trappers say that they would willingly, were it possible, make an agreement with them not to molest each other. I knew a bear to charge through an encampment of hunters who had fired upon him, knock down two of them in his progress, bite each of them slightly, and continue his flight, with no other demonstrations of anger at the warm reception he met with.

John Gray, a herculean trapper, has fought several duels with them, in which he has thus far been victorious, though generally at the expense of a gun, which he usually manages to break in the conflict. A Shawnee Indian in the Rocky Mountains, has acquired much fame by attacking and destroying numbers of them, with an old rusty sword, which he flourishes about their ears with no little dexterity and effect.

When they are attacked in a thicket, they charge out into the prairie after their pursuers, and return back, and continue to do so when any of their assailants come near them. Hunting and chasing them on horseback is a favorite amusement of both whites and Indians, and is attended with no great danger, for a good horse will easily avoid and outstrip them; but daring hunters, by charging too close upon them, have had their horses caught and frightfully lacerated before they could extricate them, which is only effected by leaving portions of their bodies in the claws and teeth of the bear.

W.A.F.

Supplementary Articles

Number 1 - Western Literary Messenger, July 20, 1842

Chanion of the Colorado

Extract from an unpublished work, entitled "LIFE IN THE ROCKY MOUNTAINS."

The Colorado a short distance below the junction of Green and Grand rivers, enters the great chanion, which is a canal in many places more than a thousand feet deep, and bounded on either side by perpendicular walls of rock, that bid defiance to horsemen, who would descend to the river; in fact, they are seldom accessible to footmen. From the summit of the walls, a succession of rocky cedar-covered hills, and sandy plains, appear losing themselves in the distance. This chanion confines the river between two and three hundred miles; and even to those, who have seen and for years been familiar with the mightiest productions of nature, presents a scene from which they recoil with terror. Reader have you ever stood on the brink of the precipice, midway between the falls of Niagara and the Whirlpool, and looked down with terror upon the rushing waters beneath you? if so, fancy yourself there at this moment; but suppose the walls to be six hundred feet higher, and in the same proportion, more distant from each other. Let the foaming torrent beneath you vanish, and imagine a beautiful meadow, perfectly level, extending from wall to wall; through this let a calm, unruffled, yet muddy streamlet wind its way; let its borders be decked with a few small shrubs, and dwarf trees. Gaze upon the frowning hills, and burning sands, with which you are to suppose yourself, half surrounded; and you will certainly attempt to descend to the lovely scene beneath, and perhaps may resolve to step over the brook, and recline yourself in the shade of a cluster of willows; alas,! if you succeed, how sadly will you be disappointed. The little brook will gradually enlarge itself as you descend, until it becomes a mighty river, three hundred yards wide; the bushes will increase in size, and stature, until they become giant cotton woods; the narrow strip of meadow land between the walls, will expand into broad fields of verdure; and you will see, kind reader, in imagination, what many a half-starved trapper, has seen in reality, on the Colorado. These walls approach each other so closely, in some places, that the bottoms disappear; and the river being compressed to one fourth of its ordinary width, dashes through with inconceivable fury, like the Niagara, at the point I have chosen, and which renders its navigation impracticable. Its tributaries are likewise confined by walls, from their junctions some distance; which compels caravans, to travel the plains, far from the main river, and depend much upon chance, and rains, for water.

Number 2 - Western Literary Messenger, August 17, 1842.

Curious Indian Letter

Extract from an unpublished work, entitled "LIFE IN THE ROCKY MOUNTAINS."

Traversing the Deer-house Plains with a party of traders, and in company with a band of Flatt-head Indians, on our way to the Buffalo range, we observed one afternoon what might be called an Indian letter, and interesting from its rare novelty, and the ingenuity with which it was devised. As the mountain tribes are not less ignorant of the art of writing than other Indians, it would be difficult to conceive how it was possible for them to effect an interchange of ideas to any considerable extent without a personal interview and by other than oral communication. In this instance, however, though the information conveyed, embodied quite a number of distinct facts, with allusions to past events and present intentions, and combined with warnings, threats and boastings, it required no Solomon to read the characters, and no Daniel to interpret their meaning. This singular document with its date, signature and superscription, excited our astonishment not less from the novelty of its appearance and the skill with which it was prepared, than from the number of ideas it imparted and the unequivocal character of the information it expressed. No learned clerk with all the appliances of letter writing at command, could have couched in more intelligible phrase or told in less doubtful terms, the knowledge intended to be made known, though addressing the Royal Society itself, than had these uncultured savages communicating with another and equally unlettered tribe. But how, methinks I hear you inquire, with a smile of incredulity, could this have been effected? It was in this wise:

In the first place, a small extent of ground was smoothed and a map of the junction of three rivers drawn. Near them were then placed several little mounds, and a small square enclosure made of pointed twigs, planted close together, in the centre of which a stick considerably longer than the others was fixed upright in the ground, having a bit of rag fastened to it at top. A great many little conical heaps of earth were arranged round the enclosure, and red earth scattered profusely among them. At the entrance to the enclosure were the figures of two persons standing, one of whom had on a hat and was represented in the act of smoking. Behind him lay a small bunch of horse-hair rolled up and placed on a piece of tobacco. At the feet of the other were four little wooden pipes, and by his side a bit of dressed skin containing a few grains of powder. Near these persons were two sticks stuck in the ground so as to cross each other at right angles, a small stick was also planted in the ground at the foot of each of the two figures making an angle with the earth of forty-five degrees, and pointing towards the other.

There were also a magnitude of little figures of men clustered around them.

Eight or ten paces off were thirty little sticks painted red, lying on the ground. Bits of scarlet blankets and cloth were scattered about, and finally, seven small figures representing horsemen facing the north, were arranged at a little distance.

Such was the novel communication referred to, which would probably have puzzled all the academicians of Europe to decipher, but which the Flatt-heads were at no loss to understand, and which even we found little difficulty in comprehending. It was evidently a letter to the Flatt-heads, and had been arranged with great care for their inspection. The date, as we easily ascertained from indications not at all questionable, but impossible to describe in words, was of the day previous. The interpretation of this curious epistle may be rendered thus:

The situation, direction, etc. of the three rivers, and the mounds near them, made it at once certain that they were intended to represent the three forks of the Missouri. The little square enclosure or pen, presented a fort in miniature, the central stick and rag indicating its flag-staff and banner. The little conical mounds gave us at first sight a lively idea of an Indian village with its numerous lodges, and the red earth scattered among them, made it equally evident that the Indians composing the village were of the Blood tribe. The two figures of men, one wearing a hat, represent the Indian Chief and the white trader. The pipes are emblematic of peace and intimacy. The tobacco, horse-hair (for horses,) powder, and skin, show that such articles have been exchanged between them. This is confirmed by the two sticks forming a cross, which represents a sign understood by all the mountain tribes, made by placing the two fore fingers in such a position, and meaning "to trade." The two other sticks pointing from the feet of each to the breast of the other, indicates a sign made by pointing with the forefinger from the breast obliquely upwards, which is the Indian mode of declaring that it is "the truth." The multitude of little figures show evidently a large number of Indians in attendance; and the numerous bits of cloth, blankets, etc. are offered as incontestible proof of the abundant supplies they have at hand. The thirty small red sticks lying some paces off, represent as many Flatt-heads who were killed last spring, and the seven horsemen are the persons who have prepared this epistle, and who, proceeding northward, have now gone to their own country.

The communication was clearly intended to terrify the Flattheads, and warn them against hunting on the waters of the Missouri, and had it been expressed in words would have read somewhat as follows:

"Flatt-heads, take notice, that peace, amity and commerce have at length been established in good faith, between the whites and our tribe; that for our benefit they have erected a fort at the three forks of the Missouri, supplied with every thing necessary for trade that our comfort and safety require; that we have assembled in great numbers at the fort, where a brisk trade has been opened, and that we shall henceforth remain on the head waters of the Missouri. You will please observe that we scalped thirty of you last spring, and that we intend to serve the rest of you in the same manner. If, therefore, you consult your own interests and safety, you will not venture on our hunting grounds, but keep out of our vicinity. You may depend upon the truth of what we now tell you. Done by a party of seven Blood horsemen, now on our way home to the Forks."

The Chief of the Flatt-heads, a Brave, in the fullest, noblest sense of the word, after having deliberately examined this strange epistle, and satisfied himself of its signification, drawing his fine form up to its full manly height, and shrugging his shoulders with the air of one scorning their threats and hurling back defiance, while his eye flashed with the fire of inveterate hate and unconquerable courage, pronounced in the emphatic language of his tribe, and in a tone of voice, expressing far more than the simple words, or indeed any other form of phraseology could convey, a brief "ES WHAU!" (maybe!) and, turning upon his heel, gave his attention to the trivial affairs of encampment, as if nothing had occurred.

Number 3 - Dallas Herald, January 11, 1873.

Interesting Indians - The Flat Heads

The Flat Head Indians are, by far, the most interesting tribe of Indians in the Rocky Mountains. They number but fifty or sixty lodges; are nomadic in their habits, and range on the head of "Clark's Fork" of the Columbia. They have been from time immemorial, at war with the powerful tribe of the Black Feet, who probably outnumber them twenty fold. Being constitutionally brave, and frequently engaged in battle with their more numerous opponents, they are rapidly diminishing in numbers and will soon be compelled to abandon their wandering habits and settle down near some of the Forts, on the lower Columbia, where they can receive protection, and raise corn, beans, and stock for a livelihood. They derive their name "Galish", or Flat Head from an ancient custom of compressing the heads of their infants between boards so as to cause them to assume a different or unnatural shape, but this crush custom has long since been abandoned, and no living example can be found amongst them. These were the interesting and inoffensive Indians, Capt. Lewis found in Horse Prairie, on the head of the Jefferson, and enticed to his camp by a display of brads, looking glasses and other gew gaws that caught the admiration of these untutored children of the mountains. I have frequently heard "Old Guigneo," the Flat Head Chief, relate minutely, the circumstances that occurred in this, their first interview with the white man, and found that his relation was almost word for word the same as recorded by Capt. Lewis; the only difference being this, an old Shoshony or Snake squaw who was with Capt. Lewis, observing how much he seemed to be interested in these simple and amiable people, claimed them as Shoshoneys, her own people and the greatest rascals to be found in the whole mountain range. The Flat Heads boast that they have never killed a white man, or stolen a horse from any one of them. They are honest and religious. In the two years that I was trading with them, I never had so much as an awl blade stolen. Their religious exercises consist of singing and dancing, in which they all assemble and engage on every sabbath in the open air. Their customs are almost identical with those of the ancient Jews, with the single exception of burnt offerings or sacrifices, in which blood is displayed; they do make offerings to the great Saint, but these are always articles of beauty or use, as tobacco, blankets, beads and other ornaments. These are always deposited in places where they believe the Great Spirit temporarily resides, or visits. To one who has been familiar with their manners and customs, it would seem that in ancient times, they had had a Jewish Jesuit among them, who had instructed them in all the ancient doctrines of the Pentateauch. When any article, however trifling, is lost and found by any one, it is immediately handed to the Chief, who invariably restores it to the right owner. Their females are chaste and modest, and many of them would be considered as pretty anywhere. They are fond of dress, painting, and a display of brilliant colors, which makes the individual Indian when in full dress a very fantastic creature. Their appearance, however, when assembled on a gala day or on that of some annual festival, especially when on a march, is quite animating. I will attempt imperfectly to describe such a scene. We had broke up camp in Horse Prairie, on a lonely day precisely on the spot where Capt. Lewis had formerly met these Indians, and were marching eastward over the wide and beautiful plain in quest of water, grass and convenience to the game. Old Guigneo limped along at the head of the procession, his old wounds preventing him from mounting a horse. He was wrapped in a plain buffalo robe, and nowhere displayed any ornament. Behind him was led a proud war charger, which by the by, he never mounted, bearing his lance, shield, quiver, and tomahawk, and all the paraphernalia of war used in his days of vigor and prime of manhood. This horse was highly ornamented, covered with fringe and tassels of various brilliant colors, but the most prized ornament was several tailfeathers of the American Eagle, highly ornamented with porcupine quills of all colors and fastened securely to the tail of the noble animal. These indicated the number of the gallant deeds his splendid animal had enabled him to perform. His lance, too, had many of these feathers dangling at the outer ends like the small flags attached to lances by our chivalrous ancestors. These were also a record of the deeds of high emprise effected with the aid of this implement. Behind followed many gallant steeds decorated in a similar manner, and bearing the arms of their respective owners. Following these was a scattering crowd of packed horses, bearing their lodges, goods and children. These were driven by squaws, mounted on saddles having the front and rear two feet high. They all rode astride, and their saddles were as abundantly decorated with fringe, tassels, bells, etc. as the war chargers of their gallant husbands. On our left, a cavalcade of fantastic horsemen canter along, their long black hair rising and falling gracefully with the motion of their bodies. Beyond them a score of youths are trying the speed of their horses, and moving with the speed of the Arabs of the desert; beyond these still, are single horsemen at all distances in view; these are acting as spies to get a distant view of any approaching enemy, and to observe the situation, kind and numbers of game. On our right are seen half a dozen horsemen in hot pursuit of a wounded antelope, who runs nearly as fast on three legs as the following steeds do on four. Beyond them is a scattering herd of Buffalo running in all directions with twenty or thirty Indians after them. There, down goes one, then another, and still another, in a few minutes twenty are dead on the field; yonder is an Indian driving his Buffalo towards our line of march, he approaches nearly to Old Guineo, and down goes his Buffalo; this is done to get the assistance of his squaw to help butcher it. Yonder is another driving his Buffalo in the direction we are traveling; this is done to kill him as near camp as possible. Far beyond our line of march, is a single Indian waiving his Buffalo robe to and fro, and still beyond him are several Indians riding at half speed, a short distance back and forth, at right angles to our course; all this is to apprise us that they have discovered strangers. A strong party of Indians instantly depart at full speed to the point of observation always ready for peace or war; in a few moments a party of horsemen are seen far to the Northward galloping towards the distant mountains, but they are so far off that no attempt is made to follow them. We reach a clear and wooded stream where grass is high and plentiful; here a tripod is planted and Old Guineo's arms are suspended from it; other tripods are similarly set up and ornamented. The pack horses are quickly unloaded; in an hour the lodges are pitched and the goods transferred therein; in the meantime the braves are assembled here and there in circles, the calumet goes freely around, while in the midst of each of these assemblages, some old veteran in animated declaration is rehearsing the glorious deeds of their ancestors, and inspiring the listening youths to these illustrious examples. These mountain Indians have an unusual language of signs by which any Indian of any tribe can make himself clearly understood by any other Indian of any other tribe, although neither of them may understand a word of their different languages or tongues. These signs are made by graceful movements of the fingers, hands and arms, and are natural and expressive. These signs embrace animate and inanimate things; thought hope, light, darkness, truth, each has its sign, which is well understood as well as all other things, animate or otherwise, that is known to them. These signs are employed to give form and emphasis in their discourses, and a chieftain who is fond of displaying his oratorical powers reminds one of Ulysses or his older confederate in their animate discourses to the ancient Grecians. But these native declaimers introduce hundreds of gestures unknown to such men as Clay, Webster and Adams, that render their discourses much more effective than they would otherwise be, and fully make up for the paucity of their language, when compared with the fullness of ours in which ideas are more numerous but not more clearly expressive than these native effusion.

W.A.F.

Number 4 - Dallas Herald, January 27, 1873.

Personal Adventure

Leaving the rendezvous on Green River, our company was divided into several parties to make a spring hunt; each party choosing its own particular hunting ground. Our party, headed by Fontenelle and consisting of twenty-two men, determined to try our luck in Gray's Hole, a broken valley adjoining Pierre's Hole on the south, and separated by hills of no great elevation from the plains of Snake River. This valley was remarkable from the circumstance, that every stream, however small, was confined to cannons of cut rock, varying from twenty to thirty feet high, and as it was only here and there, that ravines could be found where a horseman could ascend and descend these cliffs, and there were many hiding places for Indians, it was always regarded as a very unsafe locality to trap in, for a trapper once hemmed in between the walls, had little or no chance to escape. Beaver houses was numerous; no Indians made their appearance, and our hunt was quite successful.

In the middle of May, when our hunt was about half over, Fontenelle came to me one day and proposed that I should take one or two men and hunt up the Flat Head Indians. Our hunt, said he, is quite a success, but could we induce the Flat Heads to meet us at the coming rendezvous, we should be able to get from them a considerable increase to our present store of furs, however, he continued, the Gros Ventre's of the Prairie, who have, for the last three years, been associated with the Arrappahoes, on the head waters of the Arkansas River, and south fork of the Platte, are now returning to their old friends and old haunts; the Peagans on the head waters of the Missouri, and we are here directly in the route, it is more than probable that a collision between us will take place, and for this reason it will not do to weaken our small force by sending off a party sufficient for self defense. You must, therefore, endeavor to accomplish this purpose by night traveling and extreme caution. On the following morning, having chosen an Irroquois and a Flat Head to accompany me, and finding them willing to attempt the enterprise, we made our little necessary arrangements, and set out, after bidding adieu to our comrades who did us the honor to declare that we should be seen no more alive! Directing our course across the hills, we soon came in view of the great plains of Snake River, and at the distance of sixty or seventy miles, in the distant mountains across the plain could be seen 'Cotas defile,' to which we directed our march and which would lead us over to Salmon River. The plain of Snake River was covered with wild sage, the only vegetable to be seen, except the cactus plant which was abundant of every variety. At noon, we reached the great river, and having forced our horses to swim over, we made a raft of loose drift wood upon which we placed our clothing and arms, and half pushing half swimming we reached the opposite shore. Here finding some short curly grass that is very neutricious we halted a couple of hours to let our horses graze; ate some of a small stock of dried Buffalo meat, and again proceeded. We now entered the region of the natural fort that I have described in a former article, the wild sage and cactus were still the only vegetation. The country was airid and waterless, and frequently covered with huge black basaltic boulders similar to that, that constitutes the natural forts. The famous Butes could be seen far to the southward and the Tetons that we had left behind us seem to increase in height as we receded from them. After a hard day's ride, we halted at dusk, made no fire, ate some of our dried meat and slept in one of the natural forts. After hobbling out our horses, we suffered some for water, which we had not seen since we left the river. Our repose was undisturbed. The next morning at daylight, we again set out proceeding as rapidly as the nature of the country would admit, in the direction of the defile. Towards evening, we began to see great numbers of Buffalo along the base of the yet distant mountains; these animals were in great commotion, running in all directions, evidently pursued by a large body of Indians; we however continued our course. Late in the evening we discovered three mounted men approaching us from the defile; they came within half a mile, and two of them halted, the other approached within two hundred yards, and examined us for several minutes with great attention, then galloped off rejoining his companions. They all disappeared in the defile we were about to enter, turning our course obliquely so as to strike the small stream that flows from the defile at the nearest point. We soon reached it, and concealing ourselves in the willows that were sufficiently plentiful along its margin, we made a fire and cooked and ate some of the flesh of a Buffalo we had killed during our day's march. As soon as it was dark we took the precaution to cover one of our horses that was a white color, with dark blankets to elude observation and started up the defile. We had every reason to believe that a large body of Indians were encamped in the defile, and consequently avoided going near the stream, but held our course as near the base of the adjacent mountain as the nature of the ground would admit, yet notwithstanding our precautions we had proceeded but three or four miles before all the dogs of a large village of Indians set out after us, filling the air with their yells. We rode for several miles at a break neck gallop and outstripped the dogs, after which resuming our ordinary gait, we continued our course until after midnight; turned loose our horses and slept on the spot. With the morning star we arose, sought our horses and resumed our journey. As soon as it was light enough to see, we found the ground covered with the traces of horsemen, but these diminished as we proceeded, and when we reached the narrows where the pass for several hundred yards is confined between walls of cut rock forty or fifty feet apart the numerous traces ceased to appear, and we began to congratulate ourselves that we should at leat escape the Indians we had left behind us. The pass opened into a large rolling plain which we entered and continued down for some distance. We now again began to find numerous traces of horsemen and footmen increasing in frequency as we proceeded. In a little time we found ourselves in the edge of a battle field; dead Indians killed and scalped, began to make their appearance and increased in numbers as we advanced; we counted sixteen in a few hundred yards. They were not Flat Heads or we should have recognized some of them, they appeared to have been killed only a few hours. Here it was certain that a bloody battle had been fought within a few hours past. It was also evident enough that we had passed one party of the beligerents during the preceding night, and that the other party was necessarily ahead of us, but who are they? Are they hostile or friendly? If the Flat Heads and Black Feet have had a fight, which is more than probable, which of the parties did we pass in the night, and who are we about to meet? These questions were agitated but not solved, we concluded that the chances of our meeting friends or foes were about equal, but we resolved to be extremely vigilant and the moment we should uncover Indians to seek instant concealment until that night, when we would approach them and ascertain who they were. These pleasant resolutions were all suddenly knocked in the head, for we had scarcely made them before a party of over a hundred Indians filed out of a ravine at the distance of three or four hundred yards, and at once came in full view of us. We saw at once that with our jaded horses escape was impossible, and immediately ascended a hill resolved to sell our lives as dearly as possible. The Iroquois and myself dismounted from our horses and quietly awaited the result. The Indian appeared sullen and by his tenacing grasp of his gun, and his look of firm resolve, I know that he intended to kill an advancing foe which was all he could hope to do before being killed himself. I presume my own feelings were similar to his; I had confidence in myself and gun, and believe I could accomplish as much as the Indian. The Flat Head pursued a different course. He rode at half speed back and forth on top of the mound, and in a loud but monotonous tone sung his death song, in which he recounted his deeds of heroism, the scalps he had taken and his utter readiness and contempt for death; he invited the advancing Indians to come and take his worthless life, but assured them that one of their number should accompany him to the bright hunting grounds. I confess that the example of this gallant Indian inspired me with similar feelings and I could plainly see that it had the same effect upon my Irroquois friend. All at once my Flat Head friend stopped his song and gazed intently at the advancing Indians for a moment and then dashed headlong, like a madman, down to them, crying 'Galish!' 'Galish!' All was instantly explained; the Indians were the very ones we were seeking. They had just started on a Buffalo hunt but on meeting us immediately returned to camp. Our jaded horses were turned loose and we were mounted on their war chargers and went in a sweeping gallop six or seven miles to their camp.

As soon as we arrived and had been sufficiently feasted, Guigneo assembled his warriors in council to learn the nature of my embassy. I was instructed to offer them a liberal present of ammunition, blankets, and tobacco if they would meet us at the rendezvous in Pierre's Hole. My arrival proved to be auspicious; the day before they had had a severe fight with the Blackfeet Indians, that lasted all day, and when the parties withdrew from the field of battle, the Flatheads had not a single load of powder and ball left for half of their number, and had the Blackfeet known this, the Flatheads would in all probability have been exterminated. The council at once determined to comply with our wishes, but old Guigueo said he would be compelled to make short and easy marches on account of the forty-odd wounded Indians in his village. Many of his braves had been killed, and some of the severely wounded were dying daily. On the following morning we set out, returning by the same route we had already followed. The wounded Indians were conveyed on litters consisting of two lodge poles fastened on either side of a packhorse with skins stretched on cross bars so as to form a bed for each of the sufferers. These lodge poles were very elastic and passed over rough ground like the springs of a carriage, yet it was a terrible journey to the poor invalids, though perhaps no worse than the carts and ambulances in which our own wounded are usually conveyed. We were three days in reaching the point where we had passed the Indian encampment in the night. The Indians had decamped, and we ascertained that they occupied about three times as much space as the Flatheads, and judged from this that they were about three times as numerous. When we entered the plains at the mouth of the defile, the hostile party could be seen encamped about three miles north of us on a small stream, but they made no demonstration. They had evidently had fight enough and were doubtless as much or more encumbered with wounded Indians than ourselves. We saw many fresh graves about their encampment and believed that their loss had exceeded that of the Flatheads. At the mouth of the defile, all the skins and available vessels were filled with water to supply the wounded whilst crossing the great arid plain. It required three days to reach the river, but some of the young Indians went in advance and brought back water from the river, keeping a sufficient supply on hand for the wounded.

On reaching the river, a stranger not acquainted with the resources of these Indians, would have wondered how so large a force was to be transported across a wide and swiftly flowing river without boats or rafts. He would soon, however, have been undeceived. All the goods and chattels belonging to a lodge were placed in it, it was then rolled up in the shape of a ball, puckered and tied at the top, a cord was fastened to each and they were launched into the river, the squaws and children mounted on these balls, and they were towed across the river by an Indian swimming over with the rope in his mouth. In twenty minutes the river was covered with these novel vessels, and in half an hour all were safely landed half a mile below on the opposite shore. Four of these balls with a platform of lodge poles would convey six or eight wounded Indians, and in one and a half hours from the time of our reaching the river, the tents were all stretched ready for the occupants on the opposite shore. Myself and a few youngsters fastened some drift logs together upon which we placed our clothing and arms and crossed as we had done a few days before. Finding the grass tolerably good, old Guigueo resolved to rest his wounded here for several days. On the following morning I set out accompanied by a half dozen young Indians for Pierre's Hole. Here we found Fontonelle and some other parties of other trappers who had already reached the place of rendezvous. Other parties reached camp daily, and a week later, and the Rocky Mountain Fur Company reached the valley and encamped about a mile from us the same time the two St. Louis companies headed by Dripps and Gubbitt reached the valley. Old Guigueo also made in with his forces and we now altogether numbered one thousand men able to bear arms. The Gros Venues of the prairie, as we anticipated, were passing northward in companies of one hundred lodges or more. One of these lodges fell in with the Rocky Mountain Fur Company but finding the white men abundantly able to defend themselves, they adopted the peaceful policy and opened a trade for fur and skins that was advantageous to both parties. A few days later the same Indians killed seven of our young men who had started for Saint Louis. This took place in Jackson Hole. Still later a party of these Indians numbering about a hundred and fifty warriors with their women and children, entered Pierre's Hole and had proceeded nearly to the middle of the valley before they discovered the extent of their danger. They however then acted with judgment and discretion.

They immediately sought shelter in a dense grove of aspen trees of which they hurriedly built a very substantial pen large enough to contain themselves and their horses; they also dug a trench around the pen on the inside sufficiently capacious to contain the whole party below the surface of the earth. This enabled them to shoot from the bottom of the pen and gave them a great advantage over their assailants. However, in one hour a thousand guns were constantly discharging at every hole in the pen. The party within it made a gallant resistance, and for a while returned bullet for bullet, but late in the evening resistance had almost ceased, and all parties returned to their several encampments. In the morning we visited the pen which was literally full of dead Indians, squaws, children and horses. It is not probable that any escaped. Two young and rather interesting girls, fourteen or fifteen years of age who were out hunting berries when the fight commenced, concealed themselves, and were found and captured the next morning. The Indians kindly offered to save their lives and treat them kindly, but they said that their friends and relations were all killed and they wished only for death. They were importunate, and earnestly begged for death on their knees until an old warrior finding that he could do nothing with them, released them from the bond of life with his tomahawk. Had any of the white men been present their lives would have been saved unless they had committed suicide, which from their despair was highly probable. The next morning we had sixteen Indians and six white men laid out for interment in our camp. Other parties suffered to the same extent. Many were wounded - some mortally. Subbett was shot in the breast and arm, Dripps had a bullet through his hat that took a lock of hair with it. The Indians, however, being most numerous, suffered the most, and Old Gurgueo had quite a company of wounded men to add to those already in this condition. Old Gurgueo was a great favorite with all white men, and received generous presents from all the various companies. In fact he left six weeks afterwards with ammunition and tobacco enough to last his braves for several years. He was also furnished with everything that could be found in any of our camps to alleviate the sufferings of his wounded followers. The old fellow was generous and grateful, and prayed earnestly to the Great Spirit to protect us in all our pursuits.

W.A.F.

Number 5 - Dallas Herald, December 14, 1872.

Animating Scene, and a Formidable Night Companion

In the early part of the fall we descended Bitter Root River, to the head of navigation. Here the Hudson Bay Company had a trading post which however, was abandoned in the winter. Here on a fine day we enjoyed quite an animating scene. A long line of Mackanaw boats loaded with furs and peltries, were proceeding down the river, the whole surface of which was covered with those light and fragile vessels, birch bark canoes, which of all known craft are, the most easily upset. One is astonished to see with what dexterity and address these topling things are managed by the squaws. We have seen an old squaw pack her canoe on her back two miles across the portage, set it in the water, pile up a wagon load of goods upon it, throw three or four children on top and getting into it, balance the whole by the motions of her body and paddle away with almost the speed of a fish. One is surprised that fear or no accidents occur. After the fleet of canoes had disappeared on their way down the river, a long line of horsemen also proceeded down the river by land intending to winter at Fort Colville, leaving behind a small part of the company assembled here in the morning. Others broke away in small parties, who sought shelter and game in nooks and ravines of adjacent mountains, there to spend a long and dreary winter. I accompanied a half dozen families thirty or forty miles up a small stream to a very small but pretty valley, where we hoped to find game abundant. During the early part of the winter, having a pretty good supply of provisions, and a young half breed Indian managed to keep a pretty good supply of venison on hand. The winter, however, was very severe; snow fell to the depth of five feet over the little surface of the valley and it was impossible to get about, except on snow shoes. Provisions began to grow scarce, and our hunting necessarily became an every day's business, I began to get about as tired of this as I had previously got of fishing. To chase, on snow shoes, half or three fourths of a day over spurs of mountains, kill a deer and pack it on your back to camp two or three miles, might do as an occasional amusement, but when necessity makes it an every day business it becomes rather tiresome. Convinced that we should not be able to supply all the camps, I volunteered to go down the stream where we knew a small party of Indians were stationed and hire one or two hunters to go up and hunt for the party left behind. With my gun, blanket, tomahawk and knife, together with a small scrip of dried venison I set out on snow shoes. My course lay down the narrow valley of the stream, now crossing Rocky Point, now through skirts of pine timber, and frequently over difficult ravines. During the day I saw many deer, but was not yet so lost to humanity as to kill them through mere wantoness. I suffered during the day for water, although I was almost constantly beside a stream; fifteen or eighteen inches of ice must be penetrated before I could hope to get water. Once I attempted to reach the water with my tomahawk but soon found that after penetrating the ice the water was still beyond my reach and I gave it up in despair. I had freely been eating dry snow all day, but snow is but a sorry substitute for water. In the evening I began to look about for some place where I could spend the night. Luckily, I found a cave sufficiently large at the enhance to admit me in a stooping posture. As soon as I entered it, I found it sufficiently capacious being two or three feet wide, as high, and lost in darkness in the interior. I now struck fire, and having supplied myself with a large quantity of pine knots which were sufficiently abundant, I then made myself a comfortable bed of branches of the balsam fir; got thoroughly warmed up, ate my supper of dried venison, and should really have felt quite comfortable had it not been for my inordinate thirst. Directly I heard, or fancied that I he'd, water rippling in the interior of the cave. Making a torch, and taking my gun along, I proceeded to explore the cave, and soon found, to my great joy, one of the most delightful fountains I had ever seen in my life; the water was exquisite, neither too hot or too cold, and beyond the reach of congelation. I drank until I was satisfied, having gone a little beyond the strict rules of temperance. As I sat near the spring, torch in hand, I either saw or imagined I saw a ball of fire in the interior of the cave beyond; perhaps, it was mere fancy. I saw but one and only one imperfect glance at that: Perhaps it was only the reflection of my torch on a drop of water. In any event I cared not as I knew no animal would venture to attack me with my means of keeping up a good fire. Returning to my fire I lay down, but that ball of fire recurring to my mind I took the precaution to place the muzzle of my gun in that direction, placed a knot or two on the fire and fell into a doze; whenever my fire burnt low the intense cold would awake me, some knots were placed upon the fire, and the dozing repeated. The cold was so intense, that the sap vessels of the pine trees were continually bursting with the reports as loud as fire arms. Towards day, finding that I had a much larger quantity of fuel than I should require, I built up a large fire and fell into a deeper slumber than I had yet enjoyed. I was awakened from this by a tremendous clatter and soon found that an enormous Grizzly Bear, who had probably watched me all night from the interior of the cave, most likely in trepidation, finding it impossible to resist his fears and danger had bolted for the mouth of the cave, determined to make his exit in spite of fire. The entrance to the cave was much too small to admit the passage of his body without great exertion. Being convinced that he could not return until I should have ample time to put my gun to his ear and blow his brains out, I could not resist the inclination to stick a fire brand to his long hair. In an instant he was enveloped in flames. The next moment I saw him flouncing and wallowing in the snow having the appearance of a swinged rat. He arose, shook himself, and eyed me as if he had half a notion to attack me and my fire; however, if such were his intentions, he soon changed his mind and galloped off in the direction of a ledge of rocks, no doubt promising himself never again to attempt to pass a night in that infernal cavern. I saw him no more, and never had the opportunity to inquire if he caught cold on this occasion. After eating some of my small store of venison, I went with a lighted torch, to my glorious spring not knowing how many more bears might be stored away in the dark recesses of the cave. I drank freely of the water I had discovered and left it with regret. I now departed from the cave and proceeded on my journey. At the end of three or four miles I began to see tracks of snow shoes, and the same evening reached an encampment of about fifteen Indians. I found all the trees in the neighborhood of the camp full of the carcases of deer that had been dressed but frozen hard as ice. I remained with them all night. The next morning I hired two Indians to go up and hunt for the party I had left alone. I went on down the stream, and passed the residue of the winter with an Indian friend, Payette, who had a small family and a well supplied wigwam. I amused myself by excursions on snow shoes, sometimes killed deer for amusement. Payette killed several Linxes during the winter; these were always fat and I found the flesh excellent.

Number 6 - Dallas Herald, December 14, 1872.

The First Salmon - Salmon Fishing

After a peaceful Buffalo hunt on the plains of Snake River, old Ginquro the Hot Head Chief proposed to lead us over to Salmon River, in order to give us a little recreation in the way of fishing. The first waters of the Salmon River that we reached, was a very small stream issuing from a pond or miniature lake, four or five feet deep and sixty or eighty yards in extent, having a fine pebbly bottom; the water was so clear and pure, that a pin could be seen at its greatest depth. In this pond was one single Salmon of unusual size; we inferred that this fish had selected that place to deposit its spawn to the exclusion of all other fish which it had probably driven away. After encamping, all hands, Indians and white men, surrounded the pond all elated with the idea of capturing the first salmon. Happening to have some knowledge of the habits of this fish I stepped down to the lower edge of the pond and placed myself on the shoal opposite to the deepest part in water a little over shoe mouth deep, and stood gun in hand, watching the appearance of the fish; presently, I saw it swimming down slowly close to the bottom, evidently uneasy at the presence of so many persons. As soon as it reached a depth of one foot, I aimed at his head and fired; a great splash of water followed, enveloping me and my gun. Before I could clear the water out of my eyes my comrades had the fish out on dry land, with a bullet hole through his head; he was a noble fish of the very largest size and in the finest condition. The next day we reached the shoals of Salmon River, and I lost all the glory I had acquired the day before in killing the first Salmon, for I found boys ten years old that would kill with sticks and stones three fish to my one, notwithstanding I had the advantage of a double barrel gun. The women and children were all engaged in this sport with sticks, stones, arrows, guns and lances; of the various instruments the lance seemed to be the most efficient. One old Indian with a lance would pin a fish to the ground, whilst three or four little boys followed him with forked sticks. Whenever he caught a fish, one of the boys who was always ready to place his hook in the gills of the fish and start for shore, the lance was then relieved, and the same thing repeated. Old Guigneo notwithstanding his age and infirmities, was on the shoals tomahawk in hand, and killed several fish. I here saw a trait in the Indian character that was very pleasing. Some of the young and active Indians observing how anxious Guigneo was to enjoy the sport, and how difficult it was for him, to get about with alacrity, over the smooth and slippery rocks managed with great address to draw several fish within reach of his tomahawk and he seldom failed to give the death blow to such as came within his reach. So many fish were killed that we made use of only those in high condition. The squaws split and smoke-dried large quantities. We were compelled every few days to move camp to other shoals, in consequence of the offensive epheria of stinking fish, and in truth we became as tired of fish and fishing as we had before been anxious to engage in the amusement. Some of us fancied that our camps had a fishy smell for months afterwards.

W.A.F.

Number 7 - Dallas Herald, November 30, 1872

Personal Adventure

While quite a youth the writer was frequently employed in raising and conveying the contents of caches (goods and furs) to the summer rendezvous or to Winter quarters. Care, caution and uniform success, elicited the praise of his employers, and induced them to give him frequent employment in expeditions of this kind. The amount of force sent was always determined by the leader of the company, but the partizans of these squads always had the privaledge of selecting his own assistants. On one occasion he was ordered to take five men and ten pack horses and raise a 'cache' of furs about three hundred miles distant. The course lay directly through the country occupied by a powerful tribe of Indians, the Aarrappahoes who were at the time hostile. The value of the pack horses and fur was about fifteen thousand dollars. Having chosen two Delawares, a Shawnee, and two Canadian Frenchmen, all men who had seen death in all its forms and could look it steadily in the face if necessary, yet men of great caution and vigilance who had all shown great presence of mind on trying occasions, the preliminaries being settled, we set out, generally halting at twelve o'clock and remaining until two in order to give our horses ample time to graze. At such times someone of the party was always stationed out on some eminence where he could watch the movements of game or the approach of Indians. At two hours before sun set we again halted and remained until dark, then saddled up our horses and proceeded ten or twelve miles after night, turning loose or hobbling our horses, we slept on the spot where we halted. This course prevented any straggling Indians who might have seen us during the day from following us to our encampments. It also gave us a great advantage of war parties, as we would necessarily see their fires when in the vicinity of our course and determine the extent of our night marches so as to avoid them, altogether. In this manner we proceeded on to the "caches," raised them and returned as far as Salt River, about half way to the main encampment. Three or four days previously, we had crossed the great lodge trail of the Appahoes, going south east into the defiles of the great chain of mountains that were nearly parallel to our course about a half a day's march to the eastward and in which are seen Long's and Pike's peaks and many other conspicuous land marks. The Arrapahoes had hunted out the country leaving little or no game behind them together with small probability of our meeting any considerable party of Indians. After passing the great trail we saw no further sign of Indians and began immediately to greatly relax our vigilance. Salt River some fifty miles west unites with Snake River forming Grand River which last unites with Green River, forming the Colorado of the West which by and by is a muddy stream like the Missouri, the Columbia River being on the contrary as clear as the Ohio. We were encamped on Salt River, Snake River was about thirty miles from our encampment a distance too great for a day's march with pack horses. However, about midway between the two rivers a span of the great mountain chain projected down into the plain, reaching quite to the trail generally followed and near its termination was a famous spring having clusters of willows sufficient for fire wood along its margin. This was a noted place for encamping both by white men and Indians. In the morning it rained heavily for half an hour, cleared off and the sun came out giving promise of a beautiful day; we leasurely packed up our horses intending to proceed as far as the spring and halt there during the following night. So soon as we crossed Salt River we found Buffalo and Antelopes in great numbers that had certainly not recently been disturbed. The strong mist arising after the shower from the hot arid plain rendered distant objects very indistinct and the great refraction caused distant Buffalo to appear as tall as pine trees. We proceeded slowly and killed several fat and fine buffalo; the choice meat was fastened to our pack horses and we proceeded on our march. All at once we discovered eighty or ninety red objects we supposed to be Antelopes far as the eye could reach ahead of us. They were running from us in the direction of the spring. We immediately halted and examined them with great care. I soon ascertained that they did not retain their places as a herd of Deer or Antelopes generally do, but the hindmost would pass up the line another dropping in his place and we soon noticed that they were continually passing each other; more than this, we could see no reason why a herd of antelopes should be running from us at so great distance. The conclusion arrived at, was that they were Indians, and we instantly comprehended their designs. They were so far off that they believed themselves undiscovered. They were running to meet the point of mountain undiscovered; once there, they would be secure from discovery or observation. They knew, from our course, that we were aiming to reach the spring. From this place of concealment they could watch our every movement. After night, when we had gone to rest, they would surround our encampment, secure our horses, and crawl up around our camp, and fire upon us according to their custom at daybreak. All this we instantly comprehended. The question now with us was how should we manage to save our pack horses and furs. After some time spent in consultation, we adopted a course that we believed we could make successful, by good management. We knew too well that if we attempted to save ourselves by changing our course, the Indians would know at once that they were discovered and give chase; that in the course of two or three hours they would certainly overtake us and we should necessarily be compelled to abandon our pack horses to save ourselves. Our course was to proceed on to the spring and encamp as if we were wholly unconscious of the proximity of an enemy; to appear as carelessly as possible; to build fires, cook our supper, keep our riding horses close about camp, so as to throw our saddles upon them and make our escape in case the Indians should attempt to make a raid upon us by daylight; but above all, to place a hawk eyed Indian behind a rock on a neighboring eminence to watch the point of mountain, distant about three hundred yards, with unremitted vigilance. All this was duly accomplished, we reached the spring whilst the sun was yet above the horizon; our packs were thrown off with the greatest apparent carelessness, yet everything was so placed that it could instantly be found and used; our pack horses were permitted to stray some distance from camp yet our riding horses were herded close at hand. We built large fires, cooked and ate as heartily as we should have done had we known ourselves perfectly free of danger, and as soon as it became sufficiently dark to prevent observation from the point of mountain, we hurriedly but securely, packed up our horses and having built our fires and placed some saddle blankets on sticks about the fires, to give them the appearance of men sitting around them. We set out giving the point of mountain a wide berth. As we passed, however, we could plainly see the Indians in a ravine by a small fire stark naked in the act of painting themselves preliminary to their contemplated attack upon us. The next morning at sunrise we were thirty miles away and had so completely out-witted the Indians that they made no attempt to follow us. In due time we reached camp without loss and received undeserved praise for our conduct; whereas we should have been censured for exposing ourselves unnecessarily, for we should with our accustomed vigilance have performed this part of our journey by night with perfect security.

W.A.F.

Number 8 - Dallas Herald, January 4, 1873

Interesting Indians - The Nabahoes

There are really no Indians in the Rocky Mountains who deserve honorable mention, but two tribes, the Nabahoes and Flat Heads; all others are rascally, beggarly, thieves and rogues generally. The Nabahoes are found in the gulches and canons of the Gila, west of Santa Fe. They are generally at war with the Mexicans, and in most instances, have proved themselves the better soldier of the two. This may in some degree, be imputed to the great national defenses in the country where they reside, consisting of inaccessible canons and precipices to which they can retire when necessary, or from which, by secret pass and outlets, they can at any time issue and attack an enemy when, perhaps, they are wholly unexpected and by these means surprise a force that may greatly exceed their own. The Mexicans have become exceedingly chary in following these Indians to their places of concealment, and in consequence, the Indians, if too weak to make a successful resistance, retired to their hiding places until their enemies depart. These Indians have sheep and cattle; they are not migratory or nomadic, but remain permanently fixed in coves and small valleys surrounded by walls of cut rock or mountains scarcely accessible. They have attained to a high degree of perfection in the art of manufacturing blankets, similar to those made in Mexico, but we are assured that they greatly surpass the Mexicans; both in execution and design. We have seen a Nabaho blanket that was worth five hundred dollars. The texture, was fine and closely woven; colors were very brilliant and the design extremely beautiful. One who has made himself familiar with the history of design as practiced by the silk weavers of Lyons, would be surprised to learn that a tribe of Indians in the Rocky Mountains, had attained to quite as high a degree of perfection in the art, as these French weavers. And why not? Have they not the whole field of nature before them to copy, and where else can we find a higher degree of beauty and perfection than is presented by the vegetable and floral world? I am told that it required two or three years to complete one of the finest blankets. I presume that these Indians if let alone would become harmless and inoffensive. They possess within themselves all that a pastoral people require, and having no knowledge of luxuries have no use for them.

W.A.F.

Verse

by Warren A. Ferris

The following lines were composed on a dreary Winter day during a surveying excursion remote from habitation or other indications of civilized life whilst the author lay in camp alone and lame where great danger existed.

Lament

Very forlorn and weary here
Alone I rest my frame
Far far from friends and Kindred dear
I lie a cripple lame
A streamlet flows beside my lair
Sheltered by lofty cane
Sought to arrest the chilling air
And turn the wintry rain
A thousand warblers cheer the wood
With ever changing lay
But still my mind in irksome mood
Is cheerful as this day
Which line the forests aspect drear
Imparts the unwelcome truth
That Autumn hastening year by year
Succeeds the way of youth
My Comrades gone and I alone
With anxious care opprest
I hear the forrests hollow moan
And start ah! fear exprest
But better reason checks the fault
A fault I blush to own
Tho' sterner spirits often halt
By sudden terror thrown
From reasons cheek when danger's near
As pallid cheeks declare
Yet this to him is short lived fear
Who seeks a warriors fare
With thoughts like these I sink to rest
Morpheus drowns my care
My soul with fleeting visions blest
I dream of fortune rare

Mountain Scenery

Where dark blue mountains towering rise
Whose craggy summits cleave the skies
Whose sides are decked with giant pines
With branch and encircling vines
There thund'ring torrents bounding far
From rock to rock dissolve in air
Or larger streams with lion rage
Burst through the barriers that encage
And sweeping onward to the plain
Deposit ruins and again
Proceed: but mask what force subdued
Their whirlwind terror now not rude
No longer they with fury hurl
Rocks from their beds in Eddies whirl
Tall pines, nor yet with deafening roar
Assail the echoing mountains hoar
But milder than the summer breeze
Flow gently winding through the trees
Or smoothly flow and softly glide
Through woodless plains and valleys wide

Made in the USA
Middletown, DE
14 December 2019